Dirichlet Forms and Markov Processes

North-Holland Mathematical Library

VOLUME 23

NORTH-HOLLAND PUBLISHING COMPANY
AMSTERDAM · OXFORD · NEW YORK

Dirichlet Forms and Markov Processes

Masatoshi FUKUSHIMA
Department of Mathematics, Osaka University, Toyonaka-shi 560, Japan

NORTH-HOLLAND PUBLISHING COMPANY
AMSTERDAM · OXFORD · NEW YORK
1980

KODANSHA LTD
TOKYO

KODANSHA SCIENTIFIC BOOKS

Library of Congress Cataloging in Publication Data

Fukushima, Massatoshi, 1935-
Dirichlet forms and Markov processes.

(North-Holland mathematical library ; v. 23)
(Kōdansha scientific books)
Based on the author's Dirikure keishiki to Marukofu katei, published in 1975.
Bibliography: p.
Includes index.
1. Markov processes. 2. Forms, Dirichlet.
I. Title.
QA274.7.F84 1980 519.2'33 79-27730
ISBN 0-444-85421-5

Published in co-edition with North-Holland Publishing Company

Distributors for Japan:

KODANSHA LTD.
12–21 Otowa 2-chome, Bunkyo-ku, Tokyo 112, Japan

Distributors outside Japan, U.S.A. and Canada:

NORTH-HOLLAND PUBLISHING COMPANY
AMSTERDAM · NEW YORK · OXFORD

Sole distributors for the U.S.A. and Canada:

ELSEVIER NORTH-HOLLAND, INC.
52 VANDERBILT AVENUE
NEW YORK, NY 10017

Printed in Japan

Preface

A long time has passed since an intimate relationship between classical potential theory and Brownian motion was revealed. In the meantime, both potential theory and the theory of Markov processes have developed rather separately. In particular, Beurling and Deny formulated in 1959, the theory of Dirichlet space—an axiomatic extension of classical Dirichlet integrals in the direction of Markovian semigroups. Almost simultaneously, Hunt and Dynkin introduced the notions of the Hunt process and the standard process on which a probabilistic potential theory was built.

This book is an attempt to unify these two theories. By unification the theory of Markov processes bears an intrinsic analytical tool of great use, while the theory of Dirichlet spaces acquires a deep probabilistic structure. Part I of this book contains an introductory and comprehensive account of the theory of Dirichlet forms, which is fully applied to the study of Markov processes in Part II. A brief summary at the beginning of each chapter and simple examples presented in many sections serve to help you in interpreting the text.

The present book is based on my book written in Japanese and published by Kinokuniya-shoten in 1975. Professor K. Ito of Gakusyuin University, Tokyo, kindly gave me an opportunity to rewrite it in English with new organization by adding new materials (§1.5 and Chapter 5, among others) and examples. I should like to express my gratitude to him.

Thanks are due to Professor J.L. Doob of the University of Illinois, Urbana-Champaign, whose work on harmonic functions with finite Dirichlet integrals made possible my first use of Dirichlet spaces in Brownian motion. First versions of some theorems in Chapters 4 and 6 were obtained under his constant encouragement while I was at the University of Illinois from 1969 to 1971.

I am grateful to Professor M.L. Silverstein of Washington University, St. Louis. The influence of his monograph, "Symmetric Markov Processes", Lecture Notes in Math., vol. 426, Springer Verlag, 1974, is apparent throughout the present volume. He was kind enough to help me in preparing the manuscript. I am also indebt to Mr. M. Hemler for his painstaking help in correcting the English in the manuscript. Finally, I wish to thank the editorial staff of Kodansha Scientific for their truly generous cooperation.

April, 1979
Osaka

Masatoshi FUKUSHIMA

Notation

X—locally compact separable Hausdorff space
m—positive Radon measure on X with Supp$[m] = X$
$C(X)$—the space of all bounded continuous functions on X
$C_0(X) = \{u \in C(X)$: Supp$[u]$ is compact$\}$
$C_\infty(X) = \{u \in C(X)$: $^\forall\varepsilon > 0$, $^\exists K$ compact, $|u(x)| < \varepsilon$, $^\forall x \in X - K\}$
$C_0^\infty(D)$—the space of all infinitely differentiable functions with compact support on an Euclidean domain D
$\mathbf{D}(u, v)$—Dirichlet integral
$\mathscr{F}$—Dirichlet space, i.e., the domain $\mathscr{D}[\mathscr{E}]$ of a Dirichlet form $\mathscr{E}$
$\mathscr{F}_e$—extended Dirichlet space
$\mathscr{A}_+$ or $\mathscr{A}^+$—non-negative functions in a function space $\mathscr{A}$
$\mathscr{A}_b$ or $\mathscr{A}^b$ —bounded functions in a function space $\mathscr{A}$
X_Δ—one point compactification of X
$\{\mathscr{F}_t^0\}_{t\in[0,\infty]}$ —minimum admissible family (of increasing σ-fields)
$\{\mathscr{F}_t\}_{t\in[0,\infty]}$ —minimum completed admissible family
$P(\Lambda|\Sigma)$—conditional probability with respect to a σ-field Σ
$\mathscr{B}(X)$—the family of all Borel subsets of X
$f \in \mathscr{B}(X)$ or $f \in \mathscr{B}$ (resp. $f \in \mathscr{B}^+(X)$ or $f \in \mathscr{B}^+$)—f is a numerical function (resp. non-negative function) on X measurable with respect to $\mathscr{B}(X)$
S—smooth measures
S_0—positive Radon measures of finite energy integrals
$S_{00} = \{\mu \in S_0$: $\mu(X) = 1$, $\|U_1\mu\|_\infty < \infty\}$

Contents

PART II. SYMMETRIC MARKOV PROCESSES

Part I
Dirichlet Forms

Part I

Dirichlet Forms

CHAPTER 1

Basic Theory of Dirichlet Forms

The Dirichlet form on an L^2-space is defined as a Markovian closed symmetric form (§1.1). The link connecting the theory of Dirichlet forms with Markov processes is in that the Markovian nature of a closed symmetric form is equivalent to the Markovian properties of the associated semigroup and resolvent on L^2 (§1.4).

A Markovian semigroup is transient if and only if the domain of the associated Dirichlet form can be extended to a function space (an extended Dirichlet space) relative to a weighted L^1-space by means of the completion with respect to the 0-order form (§1.5). Thus, the Dirichlet space in the original sense of Beurling and Deny is a special extended Dirichlet space in the present context.

Translation invariant Dirichlet forms are studied in Examples of §1.4 and §1.5.

1.1 Basic Notions

Let H be a real Hilbert space with inner product $(\; , \;)$. A non-negative definite symmetric bilinear form densely defined on H is henceforth called simply a *symmetric form* on H. That is, $\mathscr{E}$ is called a *symmetric form* on H if the following conditions hold:

($\mathscr{E}$.1) $\mathscr{E}$ is defined on $\mathscr{D}[\mathscr{E}] \times \mathscr{D}[\mathscr{E}]$ with values in R^1, $\mathscr{D}[\mathscr{E}]$ being a dense linear subspace of H,

($\mathscr{E}$.2) $\mathscr{E}(u, v) = \mathscr{E}(v, u)$, $\mathscr{E}(u + v, w) = \mathscr{E}(u, w) + \mathscr{E}(v, w)$,
$a\,\mathscr{E}(u, v) = \mathscr{E}(au, v)$, $\mathscr{E}(u, u) \geqq 0$, $u, v, w \in \mathscr{D}[\mathscr{E}]$, $a \in R^1$.

We call $\mathscr{D}[\mathscr{E}]$ the domain of $\mathscr{E}$.

The inner product $(\; , \;)$ on H is a specific symmetric form defined on the whole space H. Given a symmetric form $\mathscr{E}$ on H,

$$(1.1.1) \quad \begin{cases} \mathscr{E}_\alpha(u, v) = \mathscr{E}(u, v) + \alpha(u,v), \quad u, v \in \mathscr{D}[\mathscr{E}] \\ \mathscr{D}[\mathscr{E}_\alpha] = \mathscr{D}[\mathscr{E}] \end{cases}$$

defines a new symmetric form on H for each $\alpha > 0$. Note that the space

$\mathscr{D}[\mathscr{E}]$ is then a pre-Hilbert space with inner product $\mathscr{E}_\alpha$. Furthermore, $\mathscr{E}_\alpha$ and $\mathscr{E}_\beta$ determine equivalent metrics on $\mathscr{D}[\mathscr{E}]$ for different $\alpha,\beta>0$. If $\mathscr{D}[\mathscr{E}]$ is complete with respect to this metric, then $\mathscr{E}$ is said to be *closed.* In other words, a symmetric form $\mathscr{E}$ is said to be closed if

$$(\mathscr{E}.3)\quad u_n\in\mathscr{D}[\mathscr{E}],\ \mathscr{E}_1(u_n-u_m,u_n-u_m)\to 0,\ n,m\to\infty\quad\Rightarrow$$
$$\exists\, u\in\mathscr{D}[\mathscr{E}],\ \mathscr{E}_1(u_n-u,u_n-u)\to 0,\ n\to\infty.$$

Clearly $\mathscr{D}[\mathscr{E}]$ is then a real Hilbert space with inner product $\mathscr{E}_\alpha$ for each $\alpha>0$.

We say that a symmetric form $\mathscr{E}$ is *closable* if the following condition is satisfied:

$$(1.1.2)\quad u_n\in\mathscr{D}[\mathscr{E}],\ \mathscr{E}(u_n-u_m,u_n-u_m)\to 0,\ n,m\to\infty,\ (u_n,u_n)\to 0,$$
$$n\to\infty\Rightarrow\mathscr{E}(u_n,u_n)\to 0,\ n\to\infty.$$

Given two symmetric forms $\mathscr{E}^{(1)}$ and $\mathscr{E}^{(2)}$, $\mathscr{E}^{(2)}$ is said to be an extension of $\mathscr{E}^{(1)}$ if $\mathscr{D}[\mathscr{E}^{(1)}]\subset\mathscr{D}[\mathscr{E}^{(2)}]$ and $\mathscr{E}^{(2)}=\mathscr{E}^{(1)}$ on $\mathscr{D}[\mathscr{E}^{(1)}]\times\mathscr{D}[\mathscr{E}^{(1)}]$. A necessary and sufficient condition for a symmetric form to possess a closed extension is that the symmetric form is closable. To see this, suppose that $\mathscr{E}$ is a closable symmetric form. Denote by $\mathscr{A}$ the set of all $\mathscr{E}_1$-Cauchy sequences. Two sequences $\{u_n\}$, $\{u_n'\}\in\mathscr{A}$ are regarded to be equivalent if $\mathscr{E}_1(u_n-u_n',u_n-u_n')\to 0$, $n\to\infty$. Denote by $\mathscr{D}[\bar{\mathscr{E}}]$ the set of all equivalence classes. This procedure provides us with a closed symmetric form $\bar{\mathscr{E}}$ on H and $\bar{\mathscr{E}}$ is indeed the smallest closed extension of $\mathscr{E}$.

Problem 1.1.1. Give the proof of the last sentence.[1)]

Problem 1.1.2. Show that the following is a sufficient condition for a symmetric form $\mathscr{E}$ to be closable:

$$(1.1.3)\ u_n\in\mathscr{D}[\mathscr{E}],\ (u_n,u_n)\to 0,\ n\to\infty\Rightarrow\mathscr{E}(u_n,v)\to 0,\ n\to\infty,\ \forall v\in\mathscr{D}[\mathscr{E}].$$

We are exclusively concerned with the case that H is an L^2-space based on a nice topological space. Thus, we start with a locally compact separable Hausdorff space X and a positive Radon measure m on X such that $\mathrm{Supp}[m]=X$; i.e., m is a non-negative Borel measure on X which is finite on compact sets and strictly positive on each non-empty open set. $L^2=L^2(X;m)$ denotes the real L^2-space with inner product

$$(1.1.4)\quad (u,v)=\int_X u(x)v(x)m(dx),\quad u,v\in L^2.$$

1) Cf. T. Kato [54; Chap. VI, 4].

Let us call a symmetric form $\mathscr{E}$ on $L^2(X; m)$ *Markovian* if the following condition is satisfied:

($\mathscr{E}$.4) For each $\varepsilon > 0$, there exists a real function $\phi_\varepsilon(t)$, $t \in R^1$, such that

(1.1.5) $\phi_\varepsilon(t) = t, \quad \forall t \in [0, 1], \quad -\varepsilon \leqq \phi_\varepsilon(t) \leqq 1 + \varepsilon, \quad \forall t \in R^1$, and
$0 \leqq \phi_\varepsilon(t') - \phi_\varepsilon(t) \leqq t' - t$ whenever $t < t'$,

(1.1.6) $u \in \mathscr{D}[\mathscr{E}] \Rightarrow \phi_\varepsilon(u) \in \mathscr{D}[\mathscr{E}], \quad \mathscr{E}(\phi_\varepsilon(u), \ \phi_\varepsilon(u)) \leqq \mathscr{E}(u, \ u)$.

A *Dirichlet form* is by definition a symmetric form on $L^2(X; m)$ which is not only Markovian but also closed. This notion plays a primary role throughout the present volume.

We now state two other conditions which look stronger but simpler than the Markovian condition ($\mathscr{E}$.4). Given a symmetric form $\mathscr{E}$ on $L^2(X; m)$, we say that the *unit contraction* (resp. *every normal contraction*) *operates* on $\mathscr{E}$ if the following ($\mathscr{E}$.4)′ (resp. ($\mathscr{E}$.4)″) is satisfied:

($\mathscr{E}$.4)′ $u \in \mathscr{D}[\mathscr{E}], v = (0 \vee v) \wedge 1 \Rightarrow v \in \mathscr{D}[\mathscr{E}], \mathscr{E}(v, v) \leqq \mathscr{E}(u, u)$.

($\mathscr{E}$.4)″ $u \in \mathscr{D}[\mathscr{E}]$, v is a normal contraction of $u \Rightarrow$
$v \in \mathscr{D}[\mathscr{E}], \mathscr{E}(v, v) \leqq \mathscr{E}(u, u)$.

Here a function v is called a *normal contraction* of a function u if $|v(x) - v(y)| \leqq |u(x) - u(y)| \ \ \forall x, \ y \in X, \ \ |v(x)| \leqq |u(x)| \ \ \forall x \in X$. We call $v \in L^2(X; m)$ a normal contraction of $u \in L^2(X; m)$ if some Borel version of v is a normal contraction of some Borel version of u.

Notice the obvious implication: ($\mathscr{E}$.4)″ $\Rightarrow$ ($\mathscr{E}$.4)′ $\Rightarrow$ ($\mathscr{E}$.4). It turns out in §1.4 that these three conditions are actually equivalent if the form $\mathscr{E}$ is closed. Therefore, we could have employed simpler condition ($\mathscr{E}$.4)′ or ($\mathscr{E}$.4)″ instead of ($\mathscr{E}$.4) in the above definition of the Dirichlet form. However, when we want to examine concrete examples, the first condition ($\mathscr{E}$.4) is the easiest one to be checked, as we shall see in the next section. Furthermore, as we shall see in the next chapter, it is both practically and theoretically important to consider Markovian symmetric forms which are closable but not necessarily closed. They do not satisfy ($\mathscr{E}$.4)′ nor ($\mathscr{E}$.4)″ in general.

We finish this section with two more definitions relevant to a symmetric form $\mathscr{E}$ on $L^2(X; m)$. Since Supp$[m] = X$, any continuous function vanishing m-a.e. on X is identically zero. Hence the subspace $L^2(X; m) \cap C(X)$ of $C(X)$[1)] can be viewed as a subspace of $L^2(X; m)$. A *core* of $\mathscr{E}$ is by definition a subset C_1 of $\mathscr{D}[\mathscr{E}] \cap C_0(X)$ such that C_1 is dense in $\mathscr{D}[\mathscr{E}]$

1) $C(X)$ denotes the space of all bounded continuous functions on X. $C(X)$ is endowed with the norm $\|u\|_\infty = \sup_{x \in X} |u(x)|$, $u \in C(X)$.

with $\mathscr{E}_1$ norm and dense in $C_0(X)$[1] with uniform norm. $\mathscr{E}$ is called *regular* if

($\mathscr{E}$.5) $\mathscr{E}$ possesses a core.

It is clear that $\mathscr{E}$ is regular if and only if the space $\mathscr{D}[\mathscr{E}] \cap C_0(X)$ is a core of $\mathscr{E}$.

For $u \in L^2(X; m)$, the support $\mathrm{Supp}[u \cdot m]$ of the measure $u \cdot dm$ is simply denoted by $\mathrm{Supp}[u]$. When u happens to be continuous, $\mathrm{Supp}[u]$ is just the closure of $\{x \in X; u(x) \neq 0\}$. We say that $\mathscr{E}$ possesses the *local property* if

($\mathscr{E}$.6) $u, v \in \mathscr{D}[\mathscr{E}]$, $\mathrm{Supp}[u]$ and $\mathrm{Supp}[v]$ are disjoint compact sets $\Rightarrow$ $\mathscr{E}(u, v) = 0$.

§ 1.2 Examples

First of all, we exhibit a typical example of a Markovian symmetric form which is not closed.

Example 1.2.1 Let us consider the case where X is a domain D of the Euclidean n-space R^n. $C_0^\infty(D)$ denotes the set of all infinitely differentiable functions on D with compact support in D. Then

$$(1.2.1)\quad \mathscr{E}(u, v) = \sum_{i,j=1}^{n} \int_D \frac{\partial u(x)}{\partial x_i} \frac{\partial v(x)}{\partial x_j} \nu_{ij}(dx)$$

$$+ \int_{D \times D - d} (u(x) - u(y))(v(x) - v(y)) J(dxdy) + \int_D u(x)v(x)k(dx)$$

$$(1.2.2)\quad \mathscr{D}[\mathscr{E}] = C_0^\infty(D),$$

is a Markovian symmetric form on $L^2(D; m)$.

Here m is a positive Radon measure on D with $\mathrm{Supp}[m] = D$. $\nu_{ij}(1 \leqq i, j \leqq n)$ are Radon measures on D such that for any $\xi \in R^n$ and any compact set $K \subset D$,

$$(1.2.3)\quad \sum_{i,j=1}^{n} \xi_i \xi_j \nu_{ij}(K) \geqq 0,\ \nu_{ij}(K) = \nu_{ji}(K),\ 1 \leqq i, j \leqq n.$$

J is a positive symmetric Radon measure on the product space $D \times D$ off the diagonal d such that for any compact set K and open set D_1 with $K \subset D_1 \subset D$

$$(1.2.4)\quad \int_{K \times K - d} |x - y|^2 J(dxdy) < \infty,\quad J(K, D - D_1) < \infty.$$

k is a positive Radon measure on D.

---

1) $C_0(X) = \{u \in C(X)\colon \mathrm{Supp}[u] \text{ is compact}\}$.

The finiteness condition (1.2.4) of J is a necessary and sufficient condition that the second integral of (1.2.1) converges absolutely for any $u, v \in C_0^\infty(D)$, as we easily see by dividing the domain $D \times D - d$ of integration into a sum of the sets $D_1 \times D_1 - d$, $(D - D_1) \times D_1$, $D_1 \times (D - D_1)$ and $(D - D_1) \times (D - D_1) - d$, D_1 being a relatively compact set containing the supports of u and v.

To see that the form $\mathscr{E}$ defined by (1.2.1) and (1.2.2) is a symmetric form on $L^2(D;m)$, it suffices to show that the first term of the right hand side of (1.2.1) is non-negative definite. Divide R^n into cubes with sides parallel to the axis and of side length $\delta > 0$. Denote by $\{C^1, \ldots, C^l\}$ those cubes whose closures are contained in D. Take a point $\eta^{(k)}$ from each cube C^k and put $\dfrac{\partial u(\eta^{(k)})}{\partial x_i} = \xi_i^{(k)}$, u being any element of $C_0^\infty(D)$. Then by the Lebesgue theorem

$$\sum_{i,j=1}^{n} \int_D \frac{\partial u(x)}{\partial x_i} \frac{\partial u(x)}{\partial x_j} \nu_{ij}(dx) = \lim_{\delta \downarrow 0} \sum_{k=1}^{l(\delta)} \left(\sum_{i,j=1}^{n} \xi_i^{(k)} \xi_j^{(k)} \nu_{ij}(C^k) \right)$$

which is non-negative in view of (1.2.3).

In order to verify the Markovian property ($\mathscr{E}$.4), choose, for each $\varepsilon > 0$, an infinitely differentiable function $\phi_\varepsilon(t)$ satisfying the condition (1.1.5). Such function can easily be constructed (see Problem 1.2.1 below). Then, for any $u \in C_0^\infty(D)$, $\phi_\varepsilon(u) \in C_0^\infty(D)$ and

$$\mathscr{E}(\phi_\varepsilon(u), \phi_\varepsilon(u)) = \sum_{i,j=1}^{n} \int_D |\phi_\varepsilon'(u(x))|^2 \frac{\partial u(x)}{\partial x_i} \frac{\partial u(x)}{\partial x_j} \nu_{ij}(dx)$$

$$+ \int_{D \times D} (\phi_\varepsilon(u(x)) - \phi_\varepsilon(u(y)))^2 J(dxdy) + \int_D \phi_\varepsilon(u(x))\, k(dx),$$

which is not greater than $\mathscr{E}(u, u)$ because we know from (1.1.5) the inequalities $0 \leqq \phi_\varepsilon'(t) \leqq 1$, $|\phi_\varepsilon(t) - \phi_\varepsilon(s)| \leqq |t - s|$ and $|\phi_\varepsilon(t)| \leqq |t|$.

Problem 1.2.1. Consider a mollifier, e.g., $j(x) = \gamma \exp(-1/1 - |x|^2)$ for $|x| < 1$, $j(x) = 0$ for $|x| \geqq 1$, where γ is a positive constant to make $\int_{|x| \leq 1} j(x) dx = 1$. We put $j_\delta(x) = \delta^{-n} j(\delta^{-1} x)$ for $\delta > 0$. For any $\varepsilon > 0$, consider a function $\psi_\varepsilon(t) = ((-\varepsilon) \vee t) \wedge (1 + \varepsilon)$ on R^1 and put $\varphi_\varepsilon(t) = j_\delta * \psi_\varepsilon(t) = \int_{R^1} j_\delta(t - s) \psi_\varepsilon(s) ds$, $0 < \delta < \varepsilon$. Show that $\phi_\varepsilon(t)$ is then an infinitely differentiable function satisfying (1.1.5).

When u and $v \in C_0^\infty(D)$ have disjoint supports, (1.2.1) reduces to $\mathscr{E}(u, v) = -2 \int_{D \times D - d} u(x) v(y) J(dxdy)$. We can conclude from this that the form (1.2.1) possesses the local property ($\mathscr{E}$.6) if and only if $J = 0$.

The Markovian symmetric form (1.2.1), (1.2.2) is not closed since the domain is restricted to $C_0^\infty(D)$. However it is closable in many cases. Several conditions on the measures ν_{ij} yielding the closability of the associated forms are collected in § 2.1. We also prove in § 2.2 that (1.2.1) is the most general expression of a closable Markovian symmetric form on $L^2(D; m)$ possessing $C_0^\infty(D)$ as its core. The next two examples concern the simplest cases that the measures ν_{ij} in (1.2.1) are the Lebesgue measures up to constant factors. In these cases we can describe explicitly those domains larger than $C_0^\infty(D)$ which make the resulting forms closed. Let us denote by $L^2(D)$ the L^2-space with respect to the Lebesgue measure dx on D.

Example 1.2.2. Let $I = (r_1, r_2)$ be a one-dimensional interval: $-\infty \leqq r_1 < r_2 \leqq +\infty$. Let m and k be positive Radon measures on I with Supp$[m] = I$. We put

$$\text{(1.2.5)} \quad \mathbf{D}(u, v) = \int_{r_1}^{r_2} \frac{du(x)}{dx} \frac{dv(x)}{dx} dx$$

$$\text{(1.2.6)} \quad \mathscr{F}^R = \{u \in L^2(I;\ m) \cap L^2(I;\ k) : u \text{ is absolutely continuous and } \mathbf{D}(u, u) < \infty\}.$$

Then the form $\mathscr{E}$ defined by

$$\text{(1.2.7)} \quad \begin{cases} \mathscr{E}(u, v) = \dfrac{1}{2}\mathbf{D}(u, v) + \displaystyle\int_{r_1}^{r_2} u(x)v(x)k(dx) \\ \mathscr{D}[\mathscr{E}] = \mathscr{F}^R \end{cases}$$

is a Dirichlet form on $L^2(I; m)$.

For any $\varepsilon > 0$ and $u \in \mathscr{F}^R$, take a function $\phi_\varepsilon(t)$ of Problem 1.2.1 and make a composite function $\phi_\varepsilon(u)$. Then $\phi_\varepsilon(u) \in \mathscr{F}^R$ and

$$\mathscr{E}(\phi_\varepsilon(u), \phi_\varepsilon(u)) = \frac{1}{2}\int_I |\phi'_\varepsilon(u)|^2 \left(\frac{du}{dx}\right)^2 dx + \int_I |\phi_\varepsilon(u)|^2 k(dx) \leqq \mathscr{E}(u, u),$$

proving that $\mathscr{E}$ is Markovian.

To show the closedness, take any $\mathscr{E}_1$-Cauchy sequence $u_l \in \mathscr{F}^R$, $l = 1, 2, \cdots$. Then du_l/dx converges to some $f \in L^2(I)$ in $L^2(I)$. u_l also converges to some $u \in L^2(I; m) \cap L^2(I; k)$ in the due topology. From this and the inequality $|u(a) - u(b)|^2 \leqq |a - b|\, \mathbf{D}(u, u)$, $r_1 < a < b < r_2$, we can conclude that there is a subsequence $\{l_k\}$ such that $u_{l_k}(x)$ converges to a continuous function $\tilde{u}(x)$ uniformly on each finite closed subinterval of I. Obviously $\tilde{u} = u$ m-a.e. and k-a.e. Moreover,

$$\int_I f(x)\phi(x)dx = \lim_{l_k\to\infty} \int_I \frac{du_{l_k}(x)}{dx}\phi(x)dx$$

$$= -\lim_{l_k\to\infty} \int_I u_{l_k}(x)\phi'(x)dx = -\int_I \tilde{u}(x)\phi'(x)dx, \ \forall\phi \in C_0^\infty(I),$$

which implies that $\tilde{u}$ is absolutely continuous and $d\tilde{u}/dx = f$. Hence $\tilde{u} \in \mathscr{F}^R$ and u_l is $\mathscr{E}_1$-convergent to $\tilde{u}$.

In general, the Dirichlet form (1.2.7) is not regular unless we replace the domain $\mathscr{F}^R$ by a smaller one. Denote by $\mathscr{F}^0$ the closure of $C_0^\infty(I)$ with respect to $\mathscr{E}_1$-metric. Then the form (1.2.7) with domain restricted to $\mathscr{F}^0$ is closed and regular. According to a general theorem of § 2.1, it is also Markovian and so a regular Dirichlet form. However, its Markovian nature can be seen directly by making use of the following explicit description (1.2.9) of the space $\mathscr{F}^0$ together with the preceding function in Problem 1.2.1.

We call the left boundary point r_1 of the interval I a *regular boundary* if

$$(1.2.8) \quad -\infty < r_1, \quad m(r_1, c) + k(r_1, c) < \infty \quad (r_1 < c < r_2).$$

Regularity of the right boundary point r_2 is defined in the same way. Assume that r_i is regular. Then it is easy to see that each $u \in \mathscr{F}^R$ admits a finite boundary value $u(r_i) = \lim_{x \in I, x \to r_i} u(x)$ and, moreover, $u(r_i) = 0$ whenever $u \in \mathscr{F}^0$. We now claim that this property characterizes the space $\mathscr{F}^0$, i.e.,

$$(1.2.9) \quad \mathscr{F}^0 = \{u \in \mathscr{F}^R : u(r_i) = 0 \text{ if } r_i \text{ is regular}, i = 1, 2\}.$$

In order to establish (1.2.9), we have to analyze the structure of the orthogonal complement $\mathscr{H}_1$ of $\mathscr{F}^0$ in the Hilbert space $(\mathscr{F}^R, \mathscr{E}_1)$: $\mathscr{F}^R = \mathscr{F}^0 \oplus \mathscr{H}_1$. A function $u \in \mathscr{F}^R$ belongs to $\mathscr{H}_1$ if and only if $\mathscr{E}_1(u, \varphi) = 0$, $\forall \varphi \in C_0^\infty(I)$. This means that u is absolutely continuous, a suitable version of du/dx is of bounded variation on each compact subinterval of I, and the following equation of measures holds:

$$(1.2.10) \quad \frac{1}{2} d\frac{du}{dx} - udk = udm.$$

Any absolutely continuous function u on I satisfying the equation (1.2.10) is called *1-harmonic*. It is known that the space of all 1-harmonic functions is a two dimensional vector space and that we can take as a basis of this space a strictly decreasing positive function $h^{(1)}$ and a strictly increasing positive function $h^{(2)}$. An integration by parts gives

$$\frac{1}{2}\int_a^b \left(\frac{dh^{(i)}}{dx}\right)^2 dx + \int_a^b (h^{(i)})^2 dk + \int_a^b (h^{(i)})^2 dm$$

$$= \frac{1}{2} h^{(i)}(b) \frac{dh^{(i)}}{dx}(b) - \frac{1}{2} h^{(i)}(a) \frac{dh^{(i)}}{dx}(a), \quad i = 1, 2,$$

for almost all a, b ($r_1 < a < b < r_2$). The asymptotic behaviour of the right hand side as $(a, b) \uparrow I$ is well known and it converges to a finite limit if and only if r_i is a regular boundary.[1)] Thus, $h^{(i)} \in \mathscr{H}_1$ if and only if r_i is regular. Hence,

1) Cf. K. Ito and H. P. McKean [49; §4.6, Table 1]; *See also*, K. Ito, Stochastic Processes, Iwanami Shoten, 1957. English translation by Y. Ito, Yale University. Russian translation by A. D. Wentzell, Moscow, 1961.

(1.2.11) $\mathscr{H}_1 = \{c_1 h^{(1)} + c_2 h^{(2)} : c_i = 0 \text{ unless } r_i \text{ is regular}\}$.

This implies (1.2.9). In fact, denoting by $\mathscr{F}$ the right hand side of (1.2.9), $\mathscr{F} \ominus \mathscr{F}^0 \subset \mathscr{F} \cap \mathscr{H}_1$ which consists only of the zero function, proving $\mathscr{F} = \mathscr{F}^0$. We further know from (1.2.11) that $\mathscr{F}^R = \mathscr{F}^0$ if and only if neither r_1 nor r_2 is regular.

Example 1.2.3. Consider again a domain $D \subset R^n$ and denote by $\mathbf{D}$ and $H^1(D)$ respectively the *Dirichlet integral* and the *Sobolev space* of order 1:

$$(1.2.12) \quad \mathbf{D}(u, v) = \sum_{i=1}^n \int_D \frac{\partial u}{\partial x_i} \frac{\partial v}{\partial x_i} dx$$

$$(1.2.13) \quad H^1(D) = \{u \in L^2(D): \frac{\partial u}{\partial x_i} \in L^2(D), \quad 1 \leq i \leq n\}.$$

Here the derivatives $\partial u / \partial x_i$ are taken in the sense of Schwartz distributions.

We are concerned with the form

$$(1.2.14) \quad \mathscr{E}(u, v) = \frac{1}{2} \mathbf{D}(u, v), \quad \mathscr{D}[\mathscr{E}] = H^1(D).$$

This form $\mathscr{E}$ is sometimes indicated by $(1/2\,\mathbf{D}, H^1(D))$ for convenience. This is a Dirichlet form. Indeed, the closedness follows immediately from the completeness of $L^2(D)$ and the definition of the Schwartz distribution. The Markovian nature ($\mathscr{E}$.4) can be proved exactly in the same manner as in Example 1.2.2 with the aid of the following characterization of the space $H^1(D)$[1]: $u \in H^1(D)$ if and only if for each i $(1 \leq i \leq n)$ there is a version $u^{(i)}$ of u ($u^{(i)} = u$ a.e.) such that $u^{(i)}$ is absolutely continuous on almost all straight lines parallel to x_i-axis and the derivative $\partial u^{(i)} / \partial x_i$ in the ordinary sense belongs to $L^2(D)$.

The Dirichlet form (1.2.14) is not regular in general. Denote by $H_0^1(D)$ the closure (with respect to $\mathscr{E}_1$-metric) of $C_0^\infty(D)$. It is obvious that the form $(1/2\,\mathbf{D}, H_0^1(D))$ on $L^2(D)$ is closed and regular. According to a theorem in §2.1, it is also Markovian and hence a regular Dirichlet form. When the boundary ∂D of the domain D is smooth, the following characterization of the space $H_0^1(D)$ is known,[2] which, together with the function of Problem 1.2.1, enables us to give a direct proof of the Markovian property of the form $(1/2\,\mathbf{D}, H_0^1(D))$:

$$(1.2.15) \quad H_0^1(D) = \{u \in H^1(D) : \gamma u = 0 \text{ a.e. on } \partial D\},$$

where γu stands for the limit of $u(x)$ along the normal of ∂D. It has been known that the relation (1.2.15) still holds for a general (unsmooth) domain D if ∂D and

1) Cf. S. Mizohata [69; Chap. 2].
2) Cf. S. Mizohata [69; Chap. 3].

γu are replaced by the "Martin boundary" of D and the "fine boundary value" of u respectively.[1]

Any solution u of the equation

$$(1.2.16) \quad \frac{1}{2}\Delta u(x) = \alpha u(x), \qquad x \in D$$

is (called α-*harmonic*, where Δ denotes the Laplace operator $\sum_{i=1}^{n} \frac{\partial^2}{\partial x_i^2}$, the derivatives being understood in the ordinary sense. A 0-harmonic function is simply called harmonic. We denote by $\mathscr{H}_\alpha$ the space of all α-harmonic functions on D with finite Dirichlet integrals. $\mathscr{H}_0$ is also denoted by $\mathscr{H}$. Using Weyl's lemma [2] we can see that the Hilbert space $H^1(D)$ with inner product $\mathscr{E}_\alpha(u,v) = 1/2\,\mathbf{D}(u,v) + \alpha(u,v)$ can be decomposed into a direct sum:

$$(1.2.17) \quad H^1(D) = H_0^1(D) \oplus \mathscr{H}_\alpha, \qquad \alpha > 0.$$

The symmetric form $(1/2\mathbf{D},\mathscr{H}_\alpha)$ on $L^2(D)$ has a remarkable structure in that it is unitary equivalent to a certain Dirichlet form living on the Martin boundary. In order to illustrate this, let us consider the simplest case where the domain D is the unit disk[3] : $D = \{x \in R^2\colon |x| < 1\}$. The boundary T of D is parametrized by the real number θ: $T = \{\theta\colon 0 \leqq \theta < 2\pi\}$. We introduce a Dirichlet form $\mathbf{C}$ on $L^2(T)$ by

$$(1.2.18) \quad \begin{cases} \mathbf{C}(\phi, \psi) = \dfrac{1}{16\pi}\displaystyle\int_0^{2\pi}\!\!\int_0^{2\pi} (\phi(\theta) - \phi(\theta'))(\psi(\theta) - \psi(\theta'))\sin^{-2}\left(\dfrac{\theta - \theta'}{2}\right)d\theta d\theta' \\ \mathscr{D}[\mathbf{C}] = \{\phi \in L^2(T)\colon \quad \mathbf{C}(\phi, \phi) < \infty\}. \end{cases}$$

The integral in (1.2.18) is called the *Douglas integral*. In the next example and Example 1.4.2, we consider Dirichlet forms more general than (1.2.18).

We claim that the unitary equivalence $(1/2\,\mathbf{D}, \mathscr{H}) \underset{H}{\overset{\gamma}{\rightleftarrows}} (\mathbf{C}, \mathscr{D}[\mathbf{C}])$ holds where γ is the operation of taking the boundary limit function and H is the operation of taking the Poisson integral. More specifically, we want to show

$$(1.2.19) \quad \begin{cases} \dfrac{1}{2}\mathbf{D}(H\phi, H\psi) = \mathbf{C}(\phi, \psi) \\ \mathscr{H} = H(\mathscr{D}[\mathbf{C}]), \end{cases}$$

where

1) J. L. Doob [28].
2) Cf. K. Yosida [87].
3) Cf. Example 5.5.2 for the case of demi-space.

$$H\phi(x) = \frac{1}{2\pi}\int_0^{2\pi} \frac{1-\rho^2}{1-2\rho\cos(\theta-\theta')+\rho^2}\phi(\theta')d\theta' \tag{1.2.20}$$

for $x = \rho e^{i\theta}$, $0 \leqq \rho < 1$.

First, we note that form $\mathbf{C}$ can be expressed in terms of the Fourier coefficients $c_\nu = \frac{1}{2\pi}\int_0^{2\pi} e^{-i\nu\theta'}\phi(\theta')d\theta'$ as

$$\begin{cases} \mathbf{C}(\phi,\phi) = \pi \sum_{\nu=-\infty}^{\infty} |c_\nu|^2|\nu| \\ \mathscr{D}[\mathbf{C}] = \{\phi \in L^2(T): \sum_{\nu=-\infty}^{\infty} |c_\nu|^2|\nu| < \infty\}. \end{cases} \tag{1.2.21}$$

Indeed changing variables $\eta = \theta - \theta'$, $\xi = \theta'$ and using the expression

$\phi(\xi) = \sum_{\nu=-\infty}^{\infty} c_\nu e^{i\nu\xi}$, we get

$$\begin{aligned} \mathbf{C}(\phi,\phi) &= \frac{1}{16\pi}\int_0^{2\pi}\int_0^{2\pi} (\phi(\xi+\eta)-\phi(\xi))^2 \sin^{-2}\frac{\eta}{2}\, d\eta d\xi \\ &= \frac{1}{2}\sum_\nu |c_\nu|^2 \int_0^{2\pi} \sin^2\frac{\nu\eta}{2}\sin^{-2}\frac{\eta}{2}\, d\eta = \pi\sum_\nu |c_\nu|^2|\nu|. \end{aligned}$$

Now take $\phi \in \mathscr{D}[\mathbf{C}]$. Using Green's formula on each disk of radius ρ,

$$\frac{1}{2}\mathbf{D}(H\phi, H\phi) = \lim_{\rho\uparrow 1}\frac{1}{2}\int_0^{2\pi} H\phi(\rho e^{i\theta})\frac{\partial(H\phi)}{\partial\rho}(\rho e^{i\theta})\rho d\theta.$$

On the other hand, the formula $\frac{1-\rho^2}{1-2\rho\cos(\theta-\theta')+\rho^2} = \sum_{\nu=-\infty}^{\infty} \rho^{|\nu|}e^{i\nu(\theta-\theta')}$ leads us to the expression $H\phi(\rho e^{i\theta}) = \sum c_\nu \rho^{|\nu|}e^{i\nu\theta}$, which converges uniformly in θ and in $\rho \leqq \rho_0$, $\rho_0 < 1$ being fixed. Hence, term by term calculations are legitimated and

$$\frac{1}{2}\mathbf{D}(H\phi, H\phi) = \lim_{\rho\uparrow 1} \pi \sum_{\nu=-\infty}^{\infty} |\nu|\rho^{2|\nu|}|c_\nu|^2 = \mathbf{C}(\phi,\phi)$$

proving (1.2.19).

Notice that the Dirichlet forms $(1/2\,\mathbf{D},\ H^1(D))$ and $(1/2\,\mathbf{D},\ H^1_0(D))$ possess the local property but the closely related Dirichlet form (1.2.18) does not. Finally, in this section we give a rather general example of a Dirichlet form possessing no local property.

Example 1.2.4. Let X and m be as in §1.1. Fix a metric ρ on X compatible with the given topology. Suppose that we are given a kernel $j(x,E)$ on $X \times \mathscr{B}(X)$ satisfying the following two conditions:

(j.1) For any $\varepsilon > 0$, $j(x,X - U_\varepsilon(x))$ is, as a function of $x \in X$, locally integrable with respect to m. Here $U_\varepsilon(x)$ is the ε-neighbourhood of x.

(j.2) $\int_X u(x)(jv)(x)m(dx) = \int_X (ju)(x)v(x)m(dx)$ ($\leq \infty$) for any $u, v \in \mathscr{B}^+(X)$.

j then determines a positive Radon measure $J(dx,dy)$ on $X \times X - d$ (d is the diagonal set) by

$$\int_{X\times X-d} f(x, y)\, J(dxdy) = \int_X \left\{\int_X f(x, y)\, j(x,dy)\right\} m(dx), \tag{1.2.22}$$
$$f \in C_0(X \times X - d).$$

From the condition (j.2) follows the symmetry of J:

$$\int_{X\times X-d} f(x, y)\, J(dxdy) = \int_{X\times X-d} f(y, x) J(dxdy),\ f \in C_0(X \times X - d). \tag{1.2.23}$$

Put

$$\begin{cases} \mathscr{E}(u, v) = \int_{X\times X-d} (u(x) - u(y))(v(x) - v(y)) J(dxdy) \\ \mathscr{D}[\mathscr{E}] = \{u \in L^2(X;m) \colon u \text{ is Borel measurable, } \mathscr{E}(u,u) < \infty\}. \end{cases} \tag{1.2.24}$$

Then $\mathscr{E}$ is a Dirichlet form on $L^2(X; m)$.

We first show that for a Borel function u,

$$u = 0 \quad m\text{-a.e.} \Rightarrow \mathscr{E}(u,u) = 0. \tag{1.2.25}$$

Suppose $u = 0$ m-a.e. Put $\Gamma_{K,\varepsilon} = \{(x,y) \in K \times K\colon \rho(x,y) > \varepsilon\}$ for $\varepsilon > 0$ and K compact. Then

$$\int_{\Gamma_{K,\varepsilon}} (u(x) - u(y))^2 J(dxdy) \leq 2\int_{\Gamma_{K,\varepsilon}} (u(x)^2 + u(y)^2) J(dxdy)$$
$$= 4\int_{\Gamma_{K,\varepsilon}} u(x)^2\, J(dxdy) \leq 4\int_K u(x)^2 j(x, X - U_\varepsilon(x)) m(dx) = 0.$$

By letting $\varepsilon \downarrow 0$ and $K \uparrow X$, we get $\mathscr{E}(u, u) = 0$.

We can see the Markovian property ($\mathscr{E}$.4) in the same way as in Example 1.2.1. But now we can directly check the stronger condition ($\mathscr{E}$.4)$''$: every normal contraction operates on the form (1.2.24). To prove the closedness, consider $u_l \in \mathscr{D}[\mathscr{E}]$ such that $\mathscr{E}_1(u_l - u_m, u_l - u_m) \to 0$, $l,m \to \infty$. Since u_l converges in L^2, there are a subsequence l_k and a set $N \in \mathscr{B}(X)$ with $m(N) = 0$ such that $u_{l_k}(x)$ converges on $X - N$. Put $\tilde{u}_{l_k}(x) = u_{l_k}(x)$ on $X - N$ and $\tilde{u}_{l_k}(x) = 0$ on N. Then $\tilde{u}_{l_k}(x)$ has a limit $u(x)$ everywhere and u_l converges to u in L^2. Moreover,
$$\mathscr{E}(u - u_m, u - u_m) = \int_{X-X-d} \lim_{l_k\to\infty} \{(u_{l_k}(x) - u_{l_k}(y)) - (u_m(x) - u_m(y))\}^2\, J(dxdy)$$

$\leq \lim_{l_k \to \infty} \mathscr{E}(u_{l_k} - u_m, u_{l_k} - u_m).$

The last term can be made arbitrarily small for sufficiently large m. Thus u_m is $\mathscr{E}_1$-convergent to $u \in \mathscr{D}[\mathscr{E}]$.

When X is a domain $D \subset R^n$, $C_0^\infty(D) \subset \mathscr{D}[\mathscr{E}]$ if and only if

(j.3) $\int_{K \times K} |x - y|^2 j(x, dy) m(dx) < \infty$ for any compact $K \subset D$.

§ 1.3 Closed Forms and Semigroups

In this section we consider only an abstract real Hilbert space H with inner product $(\, , \,)$.

Consider a family $\{T_t, t > 0\}$ of linear operators on H satisfying the following conditions:

(T_t.1) each T_t is a symmetric operator with domain $\mathscr{D}(T_t) = H$.
(T_t.2) semigroup property: $T_t T_s = T_{t+s}$, $t, s > 0$.
(T_t.3) contraction property: $(T_t u, T_t u) \leq (u, u)$, $t > 0$, $u \in H$.

Then $\{T_t, t > 0\}$ is called a *semigroup* (of symmetric operators) on H. It is called *strongly continuous* if in addition

(T_t.4) $(T_t u - u, T_t u - u) \to 0$, $t \downarrow 0$, $u \in H$.

A *resolvent* on H is by definition a family $\{G_\alpha, \alpha > 0\}$ of linear operators on H satisfying the following conditions:

(G_α.1) each G_α is a symmetric operator with domain $\mathscr{D}(G_\alpha) = H$.
(G_α.2) resolvent equation: $G_\alpha - G_\beta + (\alpha - \beta) G_\alpha G_\beta = 0$.
(G_α.3) contraction property: $(\alpha G_\alpha u, \alpha G_\alpha u) \leq (u, u)$, $\alpha > 0$, $u \in H$.

If an additional condition

(G_α.4) $(\alpha G_\alpha u - u, \alpha G_\alpha u - u) \to 0$, $\alpha \to \infty$, $u \in H$

is satisfied, the resolvent is said to be *strongly continuous*.

Problem 1.3.1. Given a strongly continuous semigroup $\{T_t, t > 0\}$ on H, show that the strong limit of the Riemann sum

(1.3.1) $$G_\alpha u = \int_0^\infty e^{-\alpha t} T_t u \, dt$$

determines a strongly continuous resolvent $\{G_\alpha, \alpha > 0\}$ on H. This is called the resolvent of the given semigroup.

The *generator* A of the strongly continuous semigroup $\{T_t, t > 0\}$ on H is defined by

(1.3.2) $$\begin{cases} Au = \lim_{t \downarrow 0} \dfrac{T_t u - u}{t} \\ \mathscr{D}(A) = \{u \in H \colon Au \text{ exists as a strong limit}\}. \end{cases}$$

Given a strongly continuous resolvent $\{G_\alpha, \alpha > 0\}$ on H, let us assume that $G_\alpha u = 0$. Then $G_\beta u = 0, \forall\beta > 0$, from $(G_\alpha.2)$, and $u = \lim_{\beta\to\infty} \beta G_\beta u = 0$ from $(G_\alpha.4)$. Hence G_α is invertible and we can set

$$(1.3.3) \quad \begin{cases} Au = \alpha u - G_\alpha^{-1} u \\ \mathscr{D}(A) = G_\alpha(H). \end{cases}$$

This operator A is easily seen to be independent of $\alpha > 0$ and is called the *generator* of the given resolvent $\{G_\alpha, \alpha > 0\}$.

Lemma 1.3.1. (i) The generator of the strongly continuous semigroup on H coincides with the generator of its resolvent.
(ii) The generator of the strongly continuous resolvent is a non-positve definite self-adjoint operator.

Proof. Proof of (ii). Let $\{G_\alpha, \alpha > 0\}$ be a strongly continuous resolvent on H. Since G_α is a symmetric operator defined on the whole space H, the inverse G_α^{-1} and hence the generator A of G_α are self-adjoint. Put $f(\alpha) = (u, G_\alpha u)$ for $u \in H$, then by the resolvent equation $(d/d\alpha)f(\alpha) = -(G_\alpha u, G_\alpha u) \leqq 0$. From $(G_\alpha.3)$, $|f(\alpha)| \leqq (1/\alpha)(u, u) \to 0$, $\alpha \to \infty$. Therefore, $f(\alpha) \geqq 0$, that is, G_α is non-negative definite. This implies the same property of $-A$ because, for $u \in \mathscr{D}(A)$, $(-Au, u) = \lim_{\alpha\downarrow 0} (-Au + \alpha u, u) = \lim_{\alpha\downarrow 0} (G_\alpha^{-1} u, u) \geqq 0$.
Proof of (i). Given a strongly continuous semigroup $\{T_t, t > 0\}$ on H, denote its resolvent by $\{G_\alpha, \alpha > 0\}$. Consider the generators A and A' of T_t and G_α respectively. Take $u \in \mathscr{D}(A')$, then $u = G_\alpha v$, $\exists v \in H$. From (1.3.1)

$$\frac{1}{t}(e^{-\alpha t}T_t u - u) = -\frac{1}{t}\int_0^t e^{-\alpha s}T_s v ds \to -v, \quad t \downarrow 0,$$

which means $u \in \mathscr{D}(A)$ and $Au = \alpha u - v = A'u$, proving $A' \subset A$. Conversely, take $u \in \mathscr{D}(A)$ and put $\lim_{t\downarrow 0} (1/t)(e^{-\alpha t}T_t u - u) = -v$, $w = u - G_\alpha v$. Then $\lim_{t\downarrow 0} (1/t)(e^{-\alpha t}T_t w - w) = 0$ and hence $(-Aw, w) + \alpha(w, w) = 0$. Since $(-Aw, w) = \lim_{t\downarrow 0} (1/t)\{(w, w) - (T_t w, w)\}$ is non-negative by $(T_t.3)$, we get $w = 0$, proving $A \subset A'$. Q.E.D.

By this lemma, we are led from semigroups to self-adjoint operators. In the following we make full use of the spectral calculus relevant to self-adjoint operators.[1] A symmetric operator S on H satisfying $\mathscr{D}(S) = H$, $S^2 = S$ is called a projection operator. A family $\{E_\lambda, -\infty < \lambda < \infty\}$ of projection operators on H is called a spectral family if $E_\lambda E_\mu = E_\lambda$, $\lambda \leq \mu$, $\lim_{\lambda' \downarrow \lambda} E_{\lambda'} u = E_\lambda u$, $\lim_{\lambda \to -\infty} E_\lambda u = 0$, $\lim_{\lambda \to \infty} E_\lambda u = u$, $u \in H$. Then $0 \leqq (E_\lambda u, u) \uparrow (u, u)$, $\lambda \uparrow \infty$, for $u \in H$. Moreover, $(E_\lambda u, v)$ is of bounded variation in λ for $u, v \in H$.

Given a spectral family $\{E_\lambda, -\infty < \lambda < \infty\}$ on H and a continuous function $\phi(\lambda)$ on R^1, a self-adjoint operator A on H can be uniquely defined by the following formula. This A is denoted by $\int_{-\infty}^{\infty} \phi(\lambda) dE_\lambda$.

$$
(1.3.4) \quad \begin{cases} (Au, v) = \int_{-\infty}^{\infty} \phi(\lambda) d(E_\lambda u, v) & \forall v \in H \\ \mathscr{D}(A) = \{u \in H: \int_{-\infty}^{\infty} \phi(\lambda)^2 d(E_\lambda u, u) < \infty\}. \end{cases}
$$

In this case, it holds that $E_\lambda A u = A E_\lambda u$, $\lambda \in R^1$, $u \in \mathscr{D}(A)$.

Conversely, given a self-adjoint operator A on H, there exists a unique spectral family $\{E_\lambda, -\infty < \lambda < \infty\}$ such that $A = \int_{-\infty}^{\infty} \lambda dE_\lambda$. This is called the spectral representation of A. If A is non-negative definite, then the corresponding spectral family satisfies $E_\lambda = 0$, $\lambda < 0$.

Let us prove a converse statement to Lemma 1.3.1. Let $-A$ be a non-negative definite self-adjoint operator on H and $-A = \int_{[0,\infty)} \lambda dE_\lambda$ be its spectral representation. For any non-negative continuous function ϕ on $[0,\infty)$, we denote the self-adjoint operator $\int_{[0,\infty)} \phi(\lambda) dE_\lambda$ by $\phi(-A)$. $\phi(-A)$ is again non-negative definite.

Lemma 1.3.2. Let $-A$ be a non-negative definite operator on H. (i) $\{T_t = \exp(tA), t > 0\}$ and $\{G_\alpha = (\alpha - A)^{-1}, \alpha > 0\}$ are a strongly continuous semigroup and a strongly continuous resolvent on H respectively. (ii) The generator of T_t of (i) coincides with A. The strongly continuous semigroup possessing A as its generator is unique. The same statement holds for the resolvent.

Proof. (i) This can be proved easily with the aid of the formula $(\phi(-A)u, \psi(-A)v) = \int_{[0,\infty)} \phi(\lambda)\psi(\lambda) d(E_\lambda u, v)$, $u \in \mathscr{D}(\phi(-A))$, $v \in \mathscr{D}(\psi(-A))$ which holds for any continuous functions ϕ and ψ on $[0,\infty)$. For instance, from $\alpha G_\alpha - I = \int_{[0,\infty)} (\alpha/(\alpha + \lambda) - 1) dE_\lambda$ follows $(\alpha G_\alpha u - u, \alpha G_\alpha u - u) = \int_{[0,\infty)} ((\alpha/\alpha + \lambda) - 1)^2 d(E_\lambda u, u) \to 0$, $\alpha \to \infty$, $u \in H$, the strong continuity of G_α.

1) Cf. K. Yosida [87; Chap. XI].

(ii) We see for $u \in H$

$$\int_{[0,\infty)} \lambda^2 d(E_\lambda G_\alpha u, G_\alpha u) = \int_{[0,\infty)} \frac{\lambda^2}{(\alpha + \lambda)^2} d(E_\lambda u, u) < \infty,$$

which means that $G_\alpha(H) \subset \mathscr{D}(A)$. Then the obvious equalities $(\alpha - A) G_\alpha u = u, u \in H$ and $G_\alpha(\alpha - A)u = u, u \in \mathscr{D}(A)$, imply that the generator of $\{G_\alpha, \alpha > 0\}$ is equal to A. On the other hand, G_α is the resolvent of T_t in the sense of (1.3.1). Hence A must also be the generator of T_t in view of Lemma 1.3.1.

Finally, let $\{G'_\alpha, \alpha > 0\}$ be any strongly continuous resolvent with generator A. Put $w = G_\alpha u - G'_\alpha u$, $u \in H$, then $(\alpha - A)w = 0$. Since $-A$ is non-negative definite, we get $w = 0$, namely, $G'_\alpha = G_\alpha$. The uniqueness concerning the strongly continuous semigroup $\{T_t, t > 0\}$ follows from the right continuity of $(T_t u, v)$ in t $(u, v \in H)$ and the uniqueness theorem of the Laplace transform Q.E.D.

The above two lemmas tell us that there are one to one correspondences as indicated in Table 1 among the family of non-positive definite self-adjoint operators on H, the family of strongly continuous semigroups, and the family of strongly continuous resolvents.

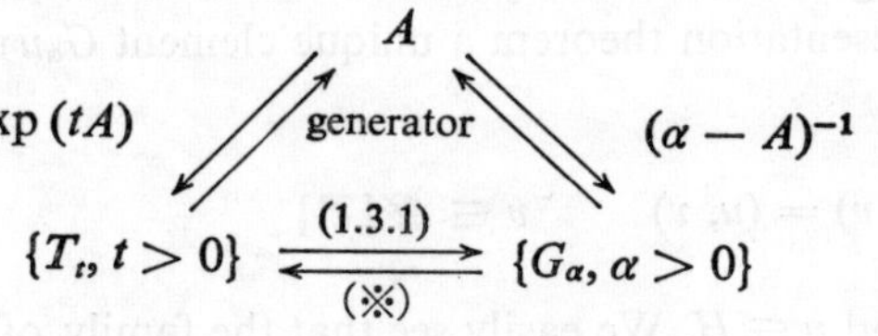

TABLE 1

Problem 1.3.2. Show that the correspondence (※) in Table 1 is given by

$$(1.3.5) \quad T_t u = \lim_{\beta \to \infty} e^{-t\beta} \sum_{n=0}^{\infty} \frac{(t\beta)^n}{n!} (\beta G_\beta)^n u, \quad u \in H.$$

We now state the main theorem of this section.

Theorem 1.3.1.[1] There is a one to one correspondence between the family of closed symmetric forms $\mathscr{E}$ on H and the family of non-positive definite self-adjoint operators A on H. The correspondence is determined by

1) Cf. T. Kato [54].

$$(1.3.6)\quad \begin{cases} \mathscr{D}[\mathscr{E}] = \mathscr{D}(\sqrt{-A}) \\ \mathscr{E}(u, v) = (\sqrt{-A}u, \ \sqrt{-A}v). \end{cases}$$

Proof. (i) Suppose $-A$ is non-negative definite and self-adjoint on H, then so is $\sqrt{-A}$. Hence $\sqrt{-A}$ is closed, which means that the symmetric form $\mathscr{E}$ on H defined by (1.3.6) is closed. Indeed, if $u_n \in \mathscr{D}(\sqrt{-A})$ satisfies $u_n \to v$, $\sqrt{-A}u_n \to w$ in H, then $v \in \mathscr{D}(\sqrt{-A})$ and $\sqrt{-A}u = w$, which is nothing but the closedness of $\mathscr{E}$.

Let us observe the following relation. The strongly continuous resolvent $\{G_\alpha, \alpha > 0\}$ generated by A satisfies

$$(1.3.7)\quad G_\alpha(H) \subset \mathscr{D}[\mathscr{E}], \ \mathscr{E}_\alpha(G_\alpha u, v) = (u, v), \ u \in H, v \in \mathscr{D}[\mathscr{E}].$$

This follows easily from the expression

$$(1.3.8)\quad \begin{cases} \mathscr{D}[\mathscr{E}] = \{u \in H: \int_{[0,\infty)} \lambda d(E_\lambda u, u) < \infty\} \\ \mathscr{E}(u, v) = \int_{[0,\infty)} \lambda d(E_\lambda u, v), \end{cases}$$

$\{E_\lambda\}$ being the spectral family associated with $-A$.

(ii) Conversely, given a closed symmetric form $\mathscr{E}$ on H, there exists by the Riesz representation theorem a unique element $G_\alpha u \in \mathscr{D}[\mathscr{E}]$ such that

$$(1.3.9)\quad \mathscr{E}_\alpha(G_\alpha u, v) = (u, v) \qquad \forall v \in \mathscr{D}[\mathscr{E}]$$

for each $\alpha > 0$ and $u \in H$. We easily see that the family of linear operators $\{G_\alpha, \alpha > 0\}$ defined this way is a strongly continuous resolvent. For instance, the contraction property follows from $\alpha(G_\alpha u, G_\alpha u) \leqq \mathscr{E}_\alpha(G_\alpha u, G_\alpha u) = (u, G_\alpha u)$ and the Schwarz inequality. To see the strong continuity, it suffices to show the strong convergence $\beta G_\beta u \to u$, $\beta \to \infty$, only for $u \in \mathscr{D}[\mathscr{E}]$ because $\mathscr{D}[\mathscr{E}]$ is dense in H and βG_β is contractive. For $u \in \mathscr{D}[\mathscr{E}]$, $\beta(\beta G_\beta u - u, \ \beta G_\beta u - u) \leqq \mathscr{E}_\beta(\beta G_\beta u - u, \ \beta G_\beta u - u) = \beta^2(G_\beta u, \ u) - \beta(u, u) + \mathscr{E}(u, u) \leqq \mathscr{E}(u, u)$, which implies the desired convergence.

Denote by A the generator of this resolvent $\{G_\alpha, \alpha > 0\}$. Since $-A$ is non-negative definite and self-adjoint, we may associate with A a closed symmetric form $\mathscr{E}'$ by the formula (1.3.6). We claim that $\mathscr{E}' = \mathscr{E}$. From (1.3.7), $G_\alpha(H) \subset \mathscr{D}[\mathscr{E}']$ and $\mathscr{E}'_\alpha(G_\alpha u, G_\alpha v) = (G_\alpha u, v)$, which also equals $\mathscr{E}_\alpha(G_\alpha u, G_\alpha v)$ by (1.3.9) for $u, v \in H$. Thus $\mathscr{E}' = \mathscr{E}$ on $G_\alpha(H) \times G_\alpha(H)$. But the same equations (1.3.7) and (1.3.9) imply that $G_\alpha(H)$ is dense both in $\mathscr{D}[\mathscr{E}']$ and in $\mathscr{D}[\mathscr{E}]$ proving $\mathscr{E}' = \mathscr{E}$.

(iii) For a given $\mathscr{E}$, A satisfying (1.3.6) is unique, because the resolvent $\{G_\alpha, \alpha > 0\}$ generated by such A satisfies the equation (1.3.7), which in turn means that $\{G_\alpha, \alpha > 0\}$ and hence A are uniquely determined by $\mathscr{E}$. Q.E.D.

In the above proof we actually showed that the correspondence between $\mathscr{E}$ and the resolvent of A is characterized by the equation (1.3.7). A restatement of this is the following.

Corollary. The correspondence in Theorem 1.3.1 can be characterized by

$$(1.3.10)\quad \begin{cases} \mathscr{D}(A) \subset \mathscr{D}[\mathscr{E}] \\ \mathscr{E}(u, v) = (-Au, v),\ u \in \mathscr{D}(A),\ v \in \mathscr{D}[\mathscr{E}]. \end{cases}$$

Example 1.3.1. Let A_0 be the self-adjoint operator on $L^2(I;m)$ corresponding to the Dirichlet form $\mathscr{E}$ of Example 1.2.2 with domain $\mathscr{F}^0$. Then

(1.3.11) $\mathscr{D}(A_0) = \{u \in L^2(I;m) \cap L^2(I;k)$: u is absolutely continuous, a version of du/dx is of bounded variation on each finite subinterval of I, $(1/2)\,(d(du/dx) - u\,dk$ is absolutely continuous with respect to m and the derivative belongs to $L^2(I;m)$, $u(r_i) = 0$ if r_i is a regular boundary$\}$.

$$(1.3.12)\quad A_0 u = \frac{\frac{1}{2}\, d\frac{du}{dx} - u\,dk}{dm}, \quad u \in \mathscr{D}(A_0).$$

We know from (1.3.10) that $u \in \mathscr{D}(A)$, $-Au = f \in L^2(I;m)$ if and only if $u \in \mathscr{D}[\mathscr{E}]$, $\mathscr{E}(u, v) = (f, v)$, $^\forall v \in \mathscr{D}[\mathscr{E}]$. Since $C_0^\infty(I)$ is dense in $\mathscr{D}[\mathscr{E}] = \mathscr{F}^0$, this is also equivalent to $u \in \mathscr{F}^0$, $\mathscr{E}(u, v) = (f, v)$, $^\forall v \in C_0^\infty(I)$, from which we get the above expression of A_0.

Exactly in the same way, we can describe the self-adjoint operator A_0 on $L^2(D)$ corresponding to the Dirichlet form $\mathscr{E}$ of Example 1.2.3 with domain $H_0^1(D)$ as follows:

$$(1.3.13)\quad \begin{cases} \mathscr{D}(A_0) = \{u \in H_0^1(D) : \Delta u \in L^2(D)\} \\ A_0 u = \frac{1}{2}\Delta u, \quad u \in \mathscr{D}(A_0). \end{cases}$$

Example 1.3.2. If we employ, instead of $\mathscr{F}^0$(resp. $H_0^1(D)$), the larger domain $\mathscr{F}^R$(resp. $H^1(D)$) for the form $\mathscr{E}$ of Example 1.2.2(resp. Example 1.2.3), then the description of the corresponding self-adjoint operator A_R becomes more delicate. To illustrate this, consider the simplest case that, in Example 1.2.2, the interval I is finite, m is the Lebesgue measure and $k = 0$. Put $\mathscr{A} = \{u \in L^2(I)$: u is differentiable, u' is absolutely continuous and $u'' \in L^2(I)\}$.

The description (1.3.11) and (1.3.12) of A_0 is now simply $\mathscr{D}(A_0) = \{u \in \mathscr{A}: u(r_1) = u(r_2) = 0\}$, $A_0 u = (1/2)u''$.

Next consider the self-adjoint operator A_R on $L^2(I)$ corresponding to the Dirichlet form 1/2 $\mathbf{D}(u, v)$ with domain $\mathscr{F}^R = \{u \in L^2(I)$: u is absolutely continuous and $\mathbf{D}(u, u) < \infty\}$. Let us show that

$$\text{(1.3.14)} \quad \begin{cases} \mathscr{D}(A_R) = \{u \in \mathscr{A}: u'(r_1) = u'(r_2) = 0\} \\ A_R u = \dfrac{1}{2} u'' \end{cases}$$

where $u'(r_i) = \lim_{x \to r_i} u'(x)$. Take any $u \in \mathscr{D}(A_R)$, then $u \in \mathscr{F}^R$ and $1/2\mathbf{D}(u, v) = (-A_R u, v)$, $\forall v \in \mathscr{F}^R$. Since $C_0^\infty(I) \subset \mathscr{F}^R$, $u \in \mathscr{A}$ and $A_R u = (1/2)u''$. Moreover, $0 = 1/2\,\mathbf{D}(u, v) + (A_R u, v) = 1/2\,\mathbf{D}(u, v) + (1/2)(u'', v) = (1/2)\{u'(r_2) \cdot v(r_2) - u'(r_1)v(r_1)\}$ for any $v \in \mathscr{F}^R$, which means that $u'(r_1) = u'(r_2) = 0$. Conversely, suppose that $u \in \mathscr{A}$ and $u'(r_1) = u'(r_2) = 0$. Then $u \in \mathscr{F}^R$ and an integration by parts leads us to $1/2\,\mathbf{D}(u, v) = (-(1/2)u'', v)$, $\forall v \in \mathscr{F}^R$. Hence $u \in \mathscr{D}(A_R)$.

The condition $u'(r_i) = 0$ in (1.3.14) is so-called boundary condition of Neumann type or of reflecting barrier. This is why we put the superscript (or subscript) R.

We state two lemmas for later use.

Lemma 1.3.3. Let a closed form $\mathscr{E}$ and a non-negative definite self-adjoint operator $-A$ correspond to each other by (1.3.6). Let $\{T_t, t > 0\}$ and $\{G_\alpha, \alpha > 0\}$ be a strongly continuous semigroup and resolvent corresponding to A. Then

(i) $T_t(H) \subset \mathscr{D}[\mathscr{E}]$, $\mathscr{E}(T_t u, T_t u) \leqq (1/2t)\{(u, u) - (T_t u, T_t u)\}$, $u \in H$.

(ii) $G_\alpha(H) \subset \mathscr{D}[\mathscr{E}]$, $\mathscr{E}_\alpha(G_\alpha u, v) = (u, v)$, $u \in H$, $v \in \mathscr{D}[\mathscr{E}]$.

(iii) The following convergence takes place strongly in $\mathscr{D}[\mathscr{E}]$ for any $u \in \mathscr{D}[\mathscr{E}]$: $T_t u \to u$, $t \downarrow 0$, $(1/t)(G_1 u - e^{-t} G_1 T_t u) = (1/t)(G_1 u - e^{-t} T_t G_1 u) \to u$, $t \downarrow 0$, $\alpha G_\alpha u \to u$, $\alpha \to \infty$.

Proof. (ii) has been shown using (1.3.8). Other assertions can be proved in the same way. By integrating $\lambda e^{-2t\lambda} \leqq (1/2t)(1 - e^{-2t\lambda})$ with $d(E_\lambda u, u)$, we get (i). For $w_t = (1/t)(G_1 u - e^{-t} G_1 T_t u)$, we have $\mathscr{E}_1(w_t - u, w_t - u) = \int_{[0,\infty)} ((1 - e^{-t(\lambda+1)})/t(\lambda + 1) - 1)^2 (\lambda + 1) d(E_\lambda u, u) \to 0$, $t \downarrow 0$, $u \in \mathscr{D}[\mathscr{E}]$. Q.E.D.

For a semigroup $\{T_t, t > 0\}$ and a resolvent $\{G_\alpha, \alpha > 0\}$ on H, we define symmetric forms $\mathscr{E}^{(t)}$ and $\mathscr{E}^{(\beta)}$ on H by

$$\text{(1.3.15)} \quad \mathscr{E}^{(t)}(u, v) = \frac{1}{t}(u - T_t u, v), \quad u, v \in H$$

(1.3.16) $\mathscr{E}^{(\beta)}(u, v) = \beta(u - \beta G_\beta u, v), \quad u, v \in H.$

The next lemma justifies our saying that $\mathscr{E}^{(t)}$ and $\mathscr{E}^{(\beta)}$ are approximating forms determined by T_t and G_β respectively.

Lemma 1.3.4. Consider $\mathscr{E}$, $-A$, T_t and G_α of Lemma 1.3.3 and let $\mathscr{E}^{(t)}$ and $\mathscr{E}^{(\beta)}$ be the approximating forms determined by T_t and G_α respectively.

(i) For $u \in H$, $\mathscr{E}^{(t)}(u, u)$ is non-decreasing as $t \downarrow 0$ and

$$(1.3.17) \quad \begin{cases} \mathscr{D}[\mathscr{E}] = \{u \in H: \lim_{t\to 0} \mathscr{E}^{(t)}(u, u) < \infty\} \\ \mathscr{E}(u, v) = \lim_{t\to 0} \mathscr{E}^{(t)}(u, v), \quad u, v \in \mathscr{D}[\mathscr{E}]. \end{cases}$$

(ii) For any $u \in H$, $\mathscr{E}^{(\beta)}(u, u)$ is non-decreasing as $\beta \uparrow \infty$ and

$$(1.3.18) \quad \begin{cases} \mathscr{D}[\mathscr{E}] = \{u \in H: \lim_{\beta\to\infty} \mathscr{E}^{(\beta)}(u, u) < \infty\} \\ \mathscr{E}(u, v) = \lim_{\beta\to\infty} \mathscr{E}^{(\beta)}(u, v), \; u, v \in \mathscr{D}[\mathscr{E}]. \end{cases}$$

This lemma can be proved using the spectral family just as in the proof of the preceding lemma. This lemma is very useful in that it provides us with a simple direct description of $\mathscr{E}$ in terms of T_t and G_α. In particular, a direct correspondence between the family of closed symmetric forms and the family of strongly continuous resolvents is given by (1.3.18) and (1.3.7).

Such a correspondence can be extended to the relationship between resolvents which are not necessarily strongly continuous and closed forms with domain which are not necessarily dense in H. We call $\mathscr{E}$ a *symmetric form* on H *in the wide sense* if $\mathscr{E}$ satisfies all conditions of the symmetric form except for the denseness of $\mathscr{D}[\mathscr{E}]$ in H.

Theorem 1.3.2 (i) Given a resolvent $\{G_\alpha, \alpha > 0\}$ on H not necessarily strongly continuous, let $\mathscr{E}^{(\beta)}$ the approximating form determined by it. Then $\mathscr{E}^{(\beta)}(u, u)$ is non-decreasing as $\beta \uparrow \infty$ for each $u \in H$. Furthermore, the form defined by the right hand side of (1.3.18) is a closed symmetric form on H in the wide sense.

(ii) Conversely, given a closed symmetric form $\mathscr{E}$ on H in the wide sense, (1.3.7) defines a resolvent $\{G_\alpha, \alpha > 0\}$ on H not necessarily strongly continuous.

(iii) The correspondence given in (i) and in (ii) are equivalent each other.

Proof. (i) The resolvent equation and the contraction property of

G_α imply that G_α is non-negative definite and $(G_\beta u, u) \leqq (1/\beta)(u, u)$. Hence $\mathscr{E}^{(\beta)}(u, u) \geqq 0$. By the resolvent equation again,

$$(1.3.19)\quad \begin{cases} \dfrac{d}{d\beta}\mathscr{E}^{(\beta)}(u, u) = (\beta G_\beta u - u, \beta G_\beta u - u) \geqq 0 \\ \dfrac{d^2}{d\beta^2}\mathscr{E}^{(\beta)}(u, u) = -2(v, G_\beta v) \leqq 0, \end{cases}$$

where $v = \beta G_\beta u - u$. We see in particular that $\mathscr{E}^{(\beta)}(u, u)$ is non-decreasing as $\beta \uparrow \infty$.

Therefore, we can define a symmetric form $\mathscr{E}$ on H in the wide sense by the right hand side of (1.3.18). In order to see that $\mathscr{E}$ is closed, assume that $u_n \in \mathscr{D}[\mathscr{E}]$ constitutes a Cauchy sequence with respect to $\mathscr{E}_1$. Then u_n converges to some element $u \in H$ in H. We have for each $\beta > 0$ $\mathscr{E}^{(\beta)}(u_n - u, u_n - u) = \lim_{m\to\infty} \mathscr{E}^{(\beta)}(u_n - u_m, u_n - u_m) \leqq \varliminf_{m\to\infty} \mathscr{E}(u_n - u_m, u_n - u_m)$, which can be made arbitrarily small uniformly in β for sufficiently large n. After letting β tend toward ∞, we see that $u \in \mathscr{D}[\mathscr{E}]$ and $\lim_{n\to\infty} \mathscr{E}(u_n - u, u_n - u) = 0$.

We note that

$$(1.3.20)\quad (\beta G_\beta u - u, \beta G_\beta u - u) \to 0,\ \beta \to \infty,\ \forall u \in \mathscr{D}[\mathscr{E}]$$

$$(1.3.21)\quad \lim_{\beta\to\infty} \mathscr{E}_\alpha^{(\beta)}(u, v) = \mathscr{E}_\alpha(u, v),\ \forall \alpha > 0,\ \forall u, v \in \mathscr{D}[\mathscr{E}],$$

where $\mathscr{E}_\alpha^{(\beta)}(u, v) = \beta(u - \beta G_{\beta+\alpha} u, v)$. (1.3.20) is a consequence of (1.3.19). (1.3.21) follows from $\mathscr{E}_\alpha^{(\beta)}(u, v) - \mathscr{E}^{(\beta)}(u, v) = \beta^2((G_\beta - G_{\beta+\alpha})u, v) = \alpha\beta^2(G_{\beta+\alpha} G_\beta u, v) = \alpha(\beta G_\beta u, \beta G_{\beta+\alpha} v)$, which converges to $\alpha(u, v)$ by virtue of (1.3.20).

The present $\{G_\alpha, \alpha > 0\}$ and $\mathscr{E}$ are related by (1.3.7). In fact, we have, for $u \in H$, $\mathscr{E}_\alpha^{(\beta)}(G_\alpha u, G_\alpha u) = \beta(G_{\beta+\alpha} u, G_\alpha u) = \beta(u, G_{\beta+\alpha} G_\alpha u) = (u, G_\alpha u) - (u, G_{\beta+\alpha} u) \to (u, G_\alpha u)$, $\beta \to \infty$, and hence $G_\alpha u \in \mathscr{D}[\mathscr{E}]$ and $\mathscr{E}_\alpha(G_\alpha u, G_\alpha u) = (u, G_\alpha u)$ in view of (1.3.21). Furthermore, (1.3.20) implies that $\mathscr{E}_\alpha(G_\alpha u, v) = \lim_{\beta\to\infty} \mathscr{E}_\alpha^{(\beta)}(G_\alpha u, v) = \lim_{\beta\to\infty} \beta(G_{\beta+\alpha} u, v) = \lim_{\beta\to\infty} (u, \beta G_{\beta+\alpha} v) = (u, v)$ for $v \in \mathscr{D}[\mathscr{E}]$.

(ii) The proof is similar to the proof (ii) of Theorem 1.3.1.

(iii) This is clear because both relations are characterized by (1.3.7). Q.E.D.

§1.4 Dirichlet forms and Markovian semigroups

Consider a topological space X and a measure m specified in §1.1.

A bounded linear operator S on $L^2(X;m)$ is called *Markovian* if $0 \leqq Su \leqq 1$ m-a.e. whenever $u \in L^2(X;m)$, $0 \leqq u \leqq 1$ m-a.e.

Theorem 1.4.1. Let $\mathscr{E}$ be a closed symmetric form on $L^2(X;m)$. Let $\{T_t, t>0\}$ and $\{G_\alpha, \alpha>0\}$ be the strongly continuous semigroup and the strongly continuous resolvent on $L^2(X;m)$ which are associated with $\mathscr{E}$ in the manner of the preceding section. Then the next five conditions are equivalent to each other:

(a) T_t is Markovian for each $t>0$.

(b) αG_α is Markovian for each $\alpha>0$.

(c) $\mathscr{E}$ is Markovian.

(d) The unit contraction operates on $\mathscr{E}$.

(e) Every normal contraction operates on $\mathscr{E}$.

A semigroup (resp. a resolvent) on $L^2(X;m)$ satisfying condition (a) (resp. (b)) is called a *Markovian semigroup* (resp. a *Markovian resolvent*). In particular, Theorem 1.4.1 means that the family of all Dirichlet forms on $L^2(X;m)$ and the family of all strongly continuous Markovian semigroups on $L^2(X;m)$ stand in one to one correspondence. The implications (a) $\Rightarrow$ (b) and (b) $\Rightarrow$ (a) are evident from (1.3.1) and (1.3.5) respectively. (e) $\Rightarrow$ (d) $\Rightarrow$ (c) is trivial. Hence it suffices to prove the relations (c) $\Rightarrow$ (b) and (b) $\Rightarrow$ (e).

Proof of (c) $\Rightarrow$ (b). Fix $\alpha>0$ and $u \in L^2(X;m)$ such that $0 \leqq u \leqq 1$ m-a.e. Introduce a quadratic form ψ on $\mathscr{D}[\mathscr{E}]$ by

$$(1.4.1)\quad \psi(v) = \mathscr{E}(v, v) + \alpha\left(v - \frac{u}{\alpha}, v - \frac{u}{\alpha}\right), \ v \in \mathscr{D}[\mathscr{E}].$$

Then, by virtue of (1.3.7), we have

$$(1.4.2)\quad \psi(G_\alpha u) + \mathscr{E}_\alpha(G_\alpha u - v, G_\alpha u - v) = \psi(v), \ v \in \mathscr{D}[\mathscr{E}].$$

In other words, $G_\alpha u$ is the unique element of $\mathscr{D}[\mathscr{E}]$ minimizing ψ. Now suppose that $\mathscr{E}$ is Markovian, i.e., there exists for each $\varepsilon>0$ a function $\phi_\varepsilon(t)$ satisfying the Markovian condition ($\mathscr{E}$.4). Put $\tilde{\phi}_\varepsilon(t) = (1/\alpha)\, \phi_{\alpha\varepsilon}(\alpha t)$ and $w = \tilde{\phi}_\varepsilon(G_\alpha u)$. Then

$$(1.4.3)\quad w \in \mathscr{D}[\mathscr{E}], \ \mathscr{E}(w, w) \leqq \mathscr{E}(G_\alpha u, G_\alpha u).$$

On the other hand, $|\tilde{\phi}_\varepsilon(t) - s| \leqq |t-s|$ for $s \in [0, 1/\alpha]$ and $t \in R^1$ and so $|w(x) - u(x)/\alpha| \leqq |G_\alpha u(x) - u(x)/\alpha|$ m-a.e. Hence $(w - u/\alpha, w - u/\alpha) \leqq (G_\alpha u - u/\alpha, G_\alpha u - u/\alpha)$. Combining this with (1.4.3), we get

$\psi(w) \leqq \psi(G_\alpha u)$ which implies that $w = G_\alpha u$. In particular $-\varepsilon \leqq G_\alpha u \leqq 1/\alpha + \varepsilon$. Since ε is arbitrary, the Markovian property of αG_α is proven. Q.E.D.

To prove the implication (b) $\Rightarrow$ (e), we need a lemma. We say that a linear operator S on $L^2(X;m)$ with $\mathscr{D}(S) = L^2(X;m)$ is a positive operator if $Su \geqq 0$ m-a.e. whenever $u \in L^2(X;m)$ and $u \geqq 0$ m-a.e.

Lemma 1.4.1.[1] (i) If S is a positive symmetric linear operator on $L^2(X;m)$, then there exists uniquely a positive symmetric Radon measure σ on the product space $X \times X$ satisfying the following property: for any Borel function $u, v \in L^2(X;m)$

$$(1.4.4) \quad (u, Sv) = \int_{X\times X} u(x)v(y)\sigma(dx,dy).$$

(ii) If in addition S is Markovian,

$$(1.4.5) \quad \sigma(X \times E) \leqq m(E) \quad \forall E \in \mathscr{B}(X).^{2)}$$

Proof. (ii) follows from (i). To prove (i), let us consider a function $f(x, y) = \sum_{i=1}^{l} u_i(x)v_i(y)$, $u_i, v_i \in C_0(X)$, on the space $X \times X$ and let

$$(1.4.6) \quad I(f) = \sum_{i=1}^{l} (u_i, Sv_i).$$

Assuming that $f(x, y) \geqq 0$, $x, y \in X$, we first show $I(f) \geqq 0$. Let $K = \bigcup_{i=1}^{l} \mathrm{Supp}\,[u_i]$. Since each u_i is uniformly continuous on the compactum K, we can choose, for any $\varepsilon > 0$, a finite decomposition $K = \sum_{k=1}^{p} E_k$, $E_k \in \mathscr{B}(X)$, and points $\xi_k \in E_k$, $1 \leqq k \leqq p$, such that $\sup_{x\in K} |u_i(x) - \tilde{u}_i(x)| < \varepsilon$, where $\tilde{u}_i(x) = \sum_{k=1}^{p} u_i(\xi_k) I_{E_k}(x)$, $1 \leqq i \leqq l$. Then $| I(f) - \sum_{i=1}^{l} (\tilde{u}_i, Sv_i)| \leqq \varepsilon \sum_{i=1}^{l} (I_K, |Sv_i|)$. On the other hand, $\sum_{i=1}^{l} (\tilde{u}_i, Sv_i) = \sum_{k=1}^{p} (I_{E_k}, Sf_{\xi_k})$ with $f_{\xi_k}(y) = f(\xi_k, y)$. The last sum is non-negative because $f \geqq 0$ and $Sf_{\xi_k} \geqq 0$ m-a.e. We have shown $I(f) \geqq 0$.

By the above observation, we conclude that (1.4.6) defines a positive

---

1) Cf. C. Dellacherie and P. A. Meyer [21; pp.;29].

2) $\mathscr{B}(X)$ denotes the family of all Borel subset of X.

linear functional on the space $\tilde{C}_0(X \times X) = \{f(x, y) = \sum_{i=1}^{l} u_i(x)\, v_i(y)$: $u_i, v_i \in C_0(X), l \geq 1\}$. The value $I(f)$ for $f \in \tilde{C}_0(X \times X)$ does not depend on the manner of the expression of f. So I can be extended to a positive linear functional on $C_0(X \times X)$. Q.E.D.

Proof of (b) $\Rightarrow$ (e). Assume αG_α is Markovian. Then by Lemma 1.4.1 there is a positive Radon measure σ_α on $X \times X$ such that

$$\text{(1.4.7)}\quad \alpha(u, G_\alpha v) = \int_{X \times X} u(x)v(y)\sigma_\alpha(dx, dy)$$

for any Borel functions $u, v \in L^2(X;m)$. Using this measure, the approximating form (1.3.16) can be rewritten as

$$\text{(1.4.8)}\quad \mathscr{E}^{(\beta)}(u, u) = \frac{1}{2}\beta \int_{X \times X} (\tilde{u}(x) - \tilde{u}(y))^2 \sigma_\beta(dx, dy) + \beta \int_X \tilde{u}(x)^2(1 - s_\beta(x))m(dx),\ u \in L^2(X; m),$$

where $s_\alpha(y) = \sigma_\alpha(X \times dy)/m(dy)$ and $\tilde{u}$ is any Borel modification of u. In view of Lemma 1.4.1,

$$\text{(1.4.9)}\quad 0 \leq s_\alpha(x) \leq 1 \ m\text{-a.e.}$$

It is evident from the expression (1.4.8) and Lemma 1.3.4 that every normal contraction operates on $\mathscr{E}$. Q.E.D.

Recall that a Dirichlet form is by definition a Markovian closed symmetric form. Thus all conditions (a) $\sim$ (e) of Theorem 1.4.1 are satisfied for a Dirichlet form. We collect below some important properties of the Dirichlet form related to the property (e).

Theorem 1.4.2 A Dirichlet form $\mathscr{E}$ on $L^2(X;m)$ possesses the following properties:

(i) $u, v \in \mathscr{D}[\mathscr{E}] \Rightarrow u \vee v, u \wedge v, u \wedge 1 \in \mathscr{D}[\mathscr{E}]$.

(ii) $u, v \in \mathscr{D}[\mathscr{E}] \cap L^\infty(X; m) \Rightarrow u \cdot v \in \mathscr{D}[\mathscr{E}]$ and $\sqrt{\mathscr{E}(u \cdot v, u \cdot v)} \leq \|u\|_\infty \sqrt{\mathscr{E}(v, v)} + \|v\|_\infty \sqrt{\mathscr{E}(u, u)}$.

(iii) $u \in \mathscr{D}[\mathscr{E}]$, $u_n = ((-n) \vee u) \wedge n \Rightarrow u_n \in \mathscr{D}[\mathscr{E}]$ and $u_n \to u$, $n \to \infty$, with respect to $\mathscr{E}_1$-metric.

(iv) $u \in \mathscr{D}[\mathscr{E}]$, $u^{(\varepsilon)} = u - ((-\varepsilon) \vee u) \wedge \varepsilon, \varepsilon > 0 \Rightarrow u^{(\varepsilon)} \in \mathscr{D}[\mathscr{E}]$ and $u^{(\varepsilon)} \to u$, $\varepsilon \downarrow 0$, with respect to $\mathscr{E}_1$-metric.

(v) $u_n, u \in \mathscr{D}[\mathscr{E}]$, $u_n \to u$, $n \to \infty$, with respect to $\mathscr{E}_1$-metric and $\phi(t)$ is a real function such that $\phi(0) = 0$, $|\phi(t) - \phi(t')| \leqq |t - t'|$ $t, t' \in R^1$ $\Rightarrow \phi(u_n), \phi(u) \in \mathscr{D}[\mathscr{E}]$ and $\phi(u_n) \to \phi(u)$, $n \to \infty$, weakly with respect to $\mathscr{E}_1$. If, in addition, $\phi(u) = u$, then the convergence is strong with respect to $\mathscr{E}_1$.

Proof. (i) By virtue of Theorem 1.4.1 (e), $u \in \mathscr{D}[\mathscr{E}]$ implies $|u| \in \mathscr{D}[\mathscr{E}]$ and $u \wedge 1 \in \mathscr{D}[\mathscr{E}]$. Then it suffices to note $u \vee v = (1/2)\{u + v + |u - v|\}$ and $u \wedge v = (1/2)\{u + v - |u - v|\}$.
(ii) By making use of formula (1.4.8), we can prove the following which is even stronger than (e): if $u_1, u_2 \in \mathscr{D}[\mathscr{E}]$, $w \in L^2(X;m)$ satisfy $|\tilde{w}(x) - \tilde{w}(y)| \leqq |\tilde{u}_1(x) - \tilde{u}_1(y)| + |\tilde{u}_2(x) - \tilde{u}_2(y)|$, $|\tilde{w}(x)| \leqq |\tilde{u}_1(x)| + |\tilde{u}_2(x)|$, $x, y \in X$, for some Borel modifications $\tilde{u}_1$, $\tilde{u}_2$ and $\tilde{w}$, then $w \in \mathscr{D}[\mathscr{E}]$ and $\sqrt{\mathscr{E}(w, w)} \leqq \sqrt{\mathscr{E}(u_1, u_1)} + \sqrt{\mathscr{E}(u_2, u_2)}$. Assertion (ii) is now obtained by setting $w = u \cdot v$, $u_1 = \|u\|_\infty \cdot v$ and $u_2 = \|v\|_\infty \cdot u$.
(iii) Since u_n is a normal contraction of u, $\mathscr{E}_1(u_n, u_n)$ is uniformly bounded by $\mathscr{E}_1(u, u)$. Moreover, $\mathscr{E}_1(u_n, G_1 v) = (u_n, v) \to (u, v) = \mathscr{E}_1(u, G_1 v)$, $n \to \infty$, $v \in L^2(X;m)$, by the formula (1.3.7). Since $G_1(L^2)$ is dense in $\mathscr{D}[\mathscr{E}]$ with metric $\mathscr{E}_1$, u_n weakly converges to u with respect to $\mathscr{E}_1$: $\mathscr{E}_1(u_n, w) \to \mathscr{E}_1(u, w)$, ${}^\forall w \in \mathscr{D}[\mathscr{E}]$. But then $\mathscr{E}_1(u_n - u, u_n - u) \leqq 2\,\mathscr{E}_1(u, u) - 2\,\mathscr{E}_1(u_n, u) \to 0$, $n \to \infty$.
(iv) The proof is the same as above because $u^{(\varepsilon)}$ is a normal contraction of u.
(v) The proof is again similar to the above since $\phi(u_n)$ is a normal contraction of u_n. Q.E.D.

As an application of Theorem 1.4.2 (iv), we state here a useful lemma concerning the regularity of the Dirichlet form.

Lemma 1.4.2. Let $\mathscr{E}$ be a Dirichlet form on $L^2(X;m)$.
(i) Suppose $\mathscr{D}[\mathscr{E}] \cap C_\infty(X)$ is both $\mathscr{E}_1$-dense in $\mathscr{D}[\mathscr{E}]$ and uniformly dense in $C_\infty(X)$.[1)] Then $\mathscr{E}$ is regular.
(ii) If $\mathscr{E}$ is regular, then for any $u \in C_0(X)$ there exist $u_n \in \mathscr{D}[\mathscr{E}] \cap C_0(X)$, $n = 1, 2, \ldots$, such that $\operatorname{Supp}[u_n] \subset \{x \in X : u(x) \neq 0\}$, $n = 1, 2, \ldots$, and u_n converges to u uniformly.

Proof. For any $u \in \mathscr{D}[\mathscr{E}]$, there exists by assumption a sequence $u_n \in \mathscr{D}[\mathscr{E}] \cap C_\infty(X)$ such that $\sqrt{\mathscr{E}_1(u - u_n, u - u_n)} \leqq 1/n$. By Theorem 1.4.2 (iv), $\sqrt{\mathscr{E}_1(u_n - \tilde{u}_n, u_n - \tilde{u}_n)} \leqq 1/n$ if we take $\tilde{u}_n = (u_n)^{(\varepsilon)}$ for suffi-

1) $C_\infty(X)$ denotes the space of all continuous functions on X vanishing at infinity, i.e., $C_\infty(X) = \{u \in C(X) : {}^\forall\varepsilon > 0, {}^\exists K \text{ compact}, |u(x)| < \varepsilon, {}^\forall x \in X - K\}$.

ciently small $\varepsilon > 0$. Then $\tilde{u}_n \in \mathscr{D}[\mathscr{E}] \cap C_0(X)$ is $\mathscr{E}_1$-convergent to u. Next assume $u \in C_0(X)$. By assumption, there exists for any $\varepsilon > 0$ a function $v \in \mathscr{D}[\mathscr{E}] \cap C_\infty(X)$ such that $\|u - v\|_\infty < \varepsilon$. Put $\tilde{v} = v^{(\varepsilon)}$. Then $\tilde{v} \in \mathscr{D}[\mathscr{E}] \cap C_0(X)$, $\mathrm{Supp}[\tilde{v}] \subset \{x \in X: u(x) \neq 0\}$ and $\|u - v\|_\infty < 2\varepsilon$. This proves not only (i) but also (ii). Q.E.D.

$\mathscr{E}$ is called a *Dirichlet form* on $L^2(X:m)$ *in the wide sense* if $\mathscr{E}$ satisfies all conditions of the Dirichlet form on $L^2(X;m)$ except for the condition that $\mathscr{D}[\mathscr{E}]$ is dense in $L^2(X;m)$. In view of the proof of Theorem 1.4.1, we see that the next statements hold.

Theorem 1.4.3 (i) The family of all Markovian resolvents on $L^2(X; m)$ not necessarily strongly continuous and the family of all Dirichlet forms in the wide sense stand in one to one correspondence in the manner of Theorem 1.3.2.
(ii) All statements of Theorem 1.4.2 remain valid for a Dirichlet form in the wide sense.

In the remainder of this section, we examine the cases that strongly continuous Markovian semigroups and resolvents on $L^2(X;m)$ are determined by Markovian transition functions and resolvent kernels respectively.

Given a measurable space $(S, \mathscr{B})$, a non-negative valued function $\kappa(x, A)$, $x \in S$, $A \in \mathscr{B}$, is called a *kernel* on $(S, \mathscr{B})$ if $\kappa(x, \cdot)$ is a positive measure on $\mathscr{B}$ for each fixed $x \in X$ and if $\kappa(\cdot, A)$ is a $\mathscr{B}$-measurable function for each fixed $A \in \mathscr{B}$. If an additional condition that $\kappa(x, S) \leq 1$, $x \in S$, is imposed, then κ is called a *Markovian kernel*. We write $\kappa u(x) = \int_S u(y)\kappa(x, dy)$ whenever the integral makes sense.

A family $\{p_t, t > 0\}$ of Markovian kernels on $(S, \mathscr{B})$ is said to be a *Markovian transition function* if

(1.4.10) $\quad p_t p_s u = p_{t+s} u \qquad t, s > 0, u \in \mathscr{B}_b$.[1]

A family $\{\alpha R_\alpha, \alpha > 0\}$ of Markovian kernels on $(S, \mathscr{B})$ is said to be a *Markovian resolvent kernel* if

(1.4.11) $\quad R_\alpha u - R_\beta u + (\alpha - \beta) R_\alpha R_\beta u = 0, \quad \alpha, \beta > 0, \quad u \in \mathscr{B}_b$.

A kernel on $(X, \mathscr{B}(X))$[2] or on $(X, \mathscr{B}^*(X))$[3] is called *m-symmetric* if

1) $u \in \mathscr{B}_b$ means that u is a founded $\mathscr{B}$-measurable function.
2) Cf. footnote 2) of pp. 24.
3) $E \in \mathscr{B}^*(X)$ means that E is universally measurable, i.e., measurable with respect to all probability measures on X.

$$(1.4.12)\quad \int_X u(x)(\kappa v)(x)m(dx) = \int_X (\kappa u)(x)v(x)m(dx) \quad (\leqq \infty)$$

for any non-negative measurable functions u and v. Assume that κ is an m-symmetric Markovian kernel. Then

$$(1.4.13)\quad \int_X (\kappa u(x))^2 m(dx) \leqq \int_X u(x)^2 m(dx), \ \forall u \in \mathscr{B}_b(x) \cap L^2(X; m).$$

In fact, by Schwarz inequality we have $(\kappa u(x))^2 \leqq \kappa 1(x) \cdot \kappa u^2(x) \leqq \kappa u^2(x)$, which leads us to (1.4.13) because of the symmetry of κ. (1.4.13) means that κ can be extended uniquely to a symmetric contractive operator K on $L^2(X;m)$. We call K a symmetric operator determined by the Markovian symmetric kernel κ.

Given an m-symmetric Markovian resolvent kernel $\{R_\alpha, \alpha > 0\}$ on $(X, \mathscr{B}(X))$ or on $(X, \mathscr{B}^*(X))$, the family of symmetric operators $\{G_\alpha, \alpha > 0\}$ on $L^2(X;m)$ determined by $\{R_\alpha, \alpha > 0\}$ is readily seen to be a Markovian resolvent on $L^2(X;m)$ not necessarily strongly continuous. By Theorem 1.4.3, there corresponds a Dirichlet form $\mathscr{E}$ on $L^2(X;m)$ in the wide sense. $\mathscr{E}$ is characterized as a closed symmetric form in the wide sense satisfying

$$(1.4.14)\quad R_1 u \in \mathscr{D}[\mathscr{E}], \ \mathscr{E}_1(R_1 u, v) = (u, v) \ \ \forall v \in \mathscr{D}[\mathscr{E}]$$

for any bounded Borel function $u \in L^2(X, m)$.

In the same way, any m-symmetric Markovian transition function determines a Markovian semigroup on $L^2(X; m)$ not necessarily strongly continuous. Let us give conditions which make them strongly continuous.

Lemma 1.4.3 (i) Let $\{p_t, t > 0\}$ be an m-symmetric Markovian transition function on $(X, \mathscr{B}(X))$ or on $(X, \mathscr{B}^*(X))$ and $\{T_t, t > 0\}$ be the associated Markovian semigroup on $L^2(X; m)$. If

$$(1.4.15)\quad \lim_{t \downarrow 0} p_t u(x) = u(x) \quad m\text{-a.e. } x \in X, \ \ \forall u \in C_0(X),$$

then $\{T_t, t > 0\}$ is strongly continuous.
(ii) Let $\{R_\alpha, \alpha > 0\}$ be an m-symmetric Markovian resolvent kernel on $(X, \mathscr{B}(X))$ or on $(X, \mathscr{B}^*(X))$ and $\{G_\alpha, \alpha > 0\}$ be the associated Markovian resolvent on $L^2(X;m)$. If

$$(1.4.16)\quad \lim_{\alpha \to \infty} \alpha R_\alpha u(x) = u(x) \quad m\text{-a.e. } x \in X, \ \ \forall u \in C_0(X),$$

then $\{G_\alpha, \alpha > 0\}$ is strongly continuous.

Proof. We only give a proof of (i). For any $u \in C_0(X)$, $p_t u$ is L^2-convergent to u because

$$\int_X (p_t u(x) - u(x))^2 m(dx) \leqq 2\int_X u(x)^2 m(dx) - 2\int_X u(x) p_t u(x) m(dx) \to 0,$$

$t \downarrow 0$. Since $C_0(X)$ is dense in $L^2(X,m)$ and T_t is contractive, $T_t u$ is L^2-convergent to u for any $u \in L^2(X;m)$. Q.E.D.

Given an m-symmetric Markovian transition function $\{p_t, t > 0\}$ satisfying the continuity condition (1.4.15), we can associate uniquely a Dirichlet form $\mathscr{E}$ on $L^2(X;m)$ according to Theorem 1.4.1 and Lemma 1.4.3. We call $\mathscr{E}$ the Dirichlet form determined by $\{p_t, t > 0\}$. In the next example, we compute the Dirichlet form determined by a translation invariant transition function on R^n.

Example 1.4.1 (Symmetric convolution semigroup on R^n).
A system of probability measures $\{\nu_t, t > 0\}$ on R^n is said to be a continuous symmetric convolution semigroup if

(1.4.17) $\nu_t * \nu_s = \nu_{t+s} \qquad t, s > 0$

(1.4.18) $\nu_t(A) = \nu_t(-A) \qquad A \in \mathscr{B}(R^n)$

(1.4.19) $\lim_{t \downarrow 0} \nu_t = \delta \qquad$ weakly,

where $\nu_t * \nu_s(A)$ denotes the convolution $\int_{R^n} \nu_t(A - y)\nu_s(dy)$ and δ is the Dirac measure concentrated at the origin. $\{\nu_t, t > 0\}$ then defines a Markovian transition function $p_t(x,A)$ on $(R^n, \mathscr{B}(R^n))$ by

(1.4.20) $p_t(x, A) = \nu_t(A - x)$

which is symmetric with respect to the Lebesgue measure in the sense of (1.4.12) and satisfies the continuity condition (1.4.15). The celebrated Lévy-Khinchin formula[1] under the present symmetry assumption reads as follows:

$$(1.4.21) \quad \begin{cases} \hat{\nu}_t(x) \left(= \int_{R^n} e^{i(x,y)} \nu_t(dy)\right) = e^{-t\psi(x)} \\ \psi(x) = \dfrac{1}{2}(Sx, x) + \int_{R^n} (1 - \cos(x, y)) J(dy), \end{cases}$$

where

(1.4.22) S is a non-negative definite $n \times n$ symmetric matrix,

1) Cf. K. Ito [47].

(1.4.23) J is a symmetric measure on $R^n - \{0\}$ such that $\int_{R^n} \frac{|x|^2}{1+|x|^2} J(dx) < \infty$.

Thus a continuous symmetric convolution semigroup $\{\nu_t, t > 0\}$ is characterized by a pair (S, J) satisfying (1.4.22) and (1.4.23) through the formula (1.4.21).

Let us now compute the Dirichlet form $\mathscr{E}$ on $L^2(R^n)$ determined by the above $\{\nu_t, t > 0\}$. We want to show

$$(1.4.24)\quad \begin{cases} \mathscr{E}(u, v) = \int_{R^n} \hat{u}(x)\bar{\hat{v}}(x)\psi(x)dx \\ \mathscr{D}[\mathscr{E}] = \{u \in L^2(R^n): \int_{R^n} |\hat{u}(x)|^2\psi(x)dx < \infty\}, \end{cases}$$

where $\hat{u}(x) = (2\pi)^{-n/2}\int_{R^n} e^{i(x,y)}u(y)dy$.

Since $\widehat{p_t u(x)} = \widehat{\nu_t^* u(x)} = \hat{\nu}_t(x)\hat{u}(x)$, we have by the Plancherel theorem,

$$\mathscr{E}^{(t)}(u, u) = \frac{1}{t}(u - T_t u, u) = \frac{1}{t}\int_{R^n} (\hat{u}(x) - \hat{\nu}_t(x)\hat{u}(x))\bar{\hat{u}}(x)\, dx$$

$$= \int_{R^n} |\hat{u}(x)|^2 \frac{1 - e^{-t\psi(x)}}{t} dx \uparrow \int_{R^n} |\hat{u}(x)|^2\psi(x)dx \quad t \downarrow 0,\ u \in L^2(R^n),$$

which proves (1.4.24).

The present Dirichlet form is regular. To see this, observe the expression $p_t u(x) = \int_{R^n} u(x+y)\nu_t(dy)$, from which we can readily conclude that p_t makes the space $C_\infty(R^n)$[1] invariant and $\|p_t u - u\|_\infty \to 0$, $t \downarrow 0$, for $u \in C_0(X)$. An analogous statement holds for the resolvent R_α of p_t:
$R_\alpha(x,A) = \int_0^\infty e^{-\alpha t}p_t(x, A)\, dt$. Since $R_\alpha(C_0)$ is $\mathscr{E}_1$-dense in $\mathscr{D}[\mathscr{E}]$, we see that $\mathscr{E}$ is regular in view of Lemma 1.4.2 (i).

When S is the unit matrix and $J = 0$, then $\psi(x) = (1/2)|x|^2$ and the Dirichlet form (1.4.24) on $L^2(R^n)$ reduces to the form $(1/2\,\mathbf{D}, H^1(R^n))$ which already appeared in Example 1.2.3. The corresponding convolution semigroup is Gaussian:

$$(1.4.25)\quad \nu_t(dy) = \frac{1}{(2\pi t)^{n/2}}\exp\left(-\frac{|y|^2}{2t}\right) dy$$

When S vanishes, the Dirichlet form (1.4.24) can be rewritten as

$$(1.4.26)\quad \begin{cases} \mathscr{E}(u, v) = \frac{1}{2}\int_{R^n\times R^n} (u(x+y) - u(x))(v(x+y) - v(x))J(dy)\, dx \\ \mathscr{D}[\mathscr{E}] = \{u \in L^2(R^n): \int_{R^n\times R^n} (u(x+y) - u(x))^2 J(dy)dx < \infty\}. \end{cases}$$

1) Cf. footnote 1) of pp. 26.

This is a special case of Example 1.2.4.

Example 1.4.2 (Symmetric convolution semigroup on the unit circle). Let Λ be the set of all real sequences $\lambda = \{\lambda_n\}_{n=-\infty}^{\infty}$ satisfying the following conditions:

(1.4.27) $\lambda_0 = 0, \quad \lambda_n = \lambda_{-n}$

(1.4.28) $\sum(\lambda_n + \lambda_m - \lambda_{n-m})\rho_n\rho_m \geqq 0$ for any real sequence $\{\rho_n\}$ with finite support.

If $\{\lambda_n\} \in \Lambda$, then $\lambda_n \geqq 0$, $\forall n$. $\lambda_n = |n|^\alpha$, $0 < \alpha \leqq 2$, $\lambda_n = \log(|n|+1)$ are elements of Λ. If $\{\lambda_n\}$ satisfies (1.4.27) and if λ_n is concave and increasing for $n > 0$, then $\{\lambda_n\} \in \Lambda$.

The notion of a continuous symmetric convolution semigroup $\{\nu_t\}$ on the unit circle $T = [0, 2\pi)$ can be defined analogously to the preceding example. Let us show that the family of all such semigroups $\{\nu_t\}_{t>0}$ and the family Λ are in one to one correspondence by the relation

(1.4.29) $\hat{\nu}_t(n) = e^{-t\lambda_n}, \qquad n = 0, \pm 1, + 2, \ldots,$

where $\hat{\nu}_t(n) = \int_0^{2\pi} e^{inx}\, \nu_t(dx)$.

On account of the Herglotz theorem[1] it suffices to prove the next assertion in order to establish the relation (1.4.29): consider a sequence $\{\lambda_n\}$ satisfying (1.4.27), then $\{\lambda_n\} \in \Lambda$ if and only if $\{\exp(-t\lambda_n)\}_{n=-\infty}^{\infty}$ is non-negative definite for every $t > 0$, namely,

(1.4.30) $\sum e^{-t\lambda_{n-m}} \rho_n\rho_m \geqq 0$

for any $\{\rho_n\}$ with finite support. If $\{\lambda_n\} \in \Lambda$, then $\sum \exp[t(\lambda_n + \lambda_m - \lambda_{n-m})]\, \rho_n\rho_m \geqq 0$ for any $\{\rho_n\}$, from which (1.4.30) follows easily. Conversely, if $\{\lambda_n\}$ satisfies (1.4.30) for all $t > 0$, then

(1.4.31) $\sum \lambda_{n-m}\rho_n\rho_m \leqq 0$ if $\{\rho_n\}$ is of finite support and $\sum \rho_n = 0$,

because the sequence $\gamma_n(t) = (1/t)(1 - e^{-t\lambda_n})$ possesses the property (1.4.31) and $\lim_{t\downarrow 0} \gamma_n(t) = \lambda_n$. Given any sequence $\{\rho_n\}$ with $\rho_0 = -\sum_{n\neq 0} \rho_n$, (1.4.31) reads $\rho_0 \sum_{n\neq 0} \lambda_n\rho_n + \rho_0 \sum_{m\neq 0} \lambda_{-m}\rho_m + \sum_{n,m\neq 0} \lambda_{n-m}\rho_n\rho_m \leqq 0$ and so $\sum_{n,m\neq 0} (\lambda_n + \lambda_m - \lambda_{n-m})\rho_n\rho_m \geqq 0$, which implies (1.4.28).

Let $\mathscr{E}$ be the Dirichlet form on $L^2(T)$ determined by a continuous symmetric convolution semigroup $\{\nu_t\}_{t>0}$ on T. Then

$$
(1.4.32)\quad \begin{cases} \mathscr{E}(u, v) = 2\pi \sum_{\nu=-\infty}^{\infty} \hat{u}(\nu)\bar{\hat{v}}(\nu)\lambda_\nu \\ \mathscr{D}[\mathscr{E}] = \{u \in L^2(T);\ \sum_{\nu=-\infty}^{\infty} |\hat{u}(\nu)|^2\lambda_\nu < \infty\}, \end{cases}
$$

1) Cf. F. Riesz and B. Sz. Nagy [76; Chap. III].

where $\{\lambda_\nu\}$ is a sequence corresponding to $\{\nu_t\}_{t>0}$ by (1.4.29) and $\hat{u}(\nu) = 1/2\pi \int_0^{2\pi} e^{-i\nu x}u(x)dx$. Indeed, using Parseval's formula, we have $(1/t)(u - T_t u, u) = (2\pi/t)\{\sum|\hat{u}(\nu)|^2 - \sum\hat{\nu}_t(\nu)|\hat{u}(\nu)|^2\} = 2\pi\sum((1-e^{-\lambda_\nu t})/t)|\hat{u}(\nu)|^2 \uparrow 2\pi\sum|\hat{u}(\nu)|^2\lambda_\nu, t\downarrow 0, u \in L^2(T)$. The Dirichlet form (1.4.32) with $\lambda_\nu = |\nu|/2$ has already appeared in Example 1.2.3.

§1.5 Transience and Extended Dirichlet Spaces

We have defined a Dirichlet form $\mathscr{E}$ as a Markovian closed symmetric form on $L^2(X;m)$. It is sometimes convenient to denote $\mathscr{D}[\mathscr{E}]$ by $\mathscr{F}$ and call $(\mathscr{F}, \mathscr{E})$ a *Dirichlet space* relative to $L^2(X;m)$. $\mathscr{F}$ is a Hilbert space with inner product $\mathscr{E}_\alpha$ for each $\alpha > 0$, but $\mathscr{F}$ is not even a pre-Hilbert space with respect to the 0-order form $\mathscr{E}$ in general. The purpose of this section is to prove that the space $(\mathscr{F}, \mathscr{E})$ can be enlarged to an "extended Dirichlet space" relative to some weighted L^1-space if and only if the associated Markovian semigroup is transient.

Let $\{T_t, t > 0\}$ be a strongly continuous Markovian semigroup on $L^2(X; m)$ and put

$$(1.5.1)\quad S_t f = \int_0^t T_s f ds, \qquad f \in L^2(X; m),$$

the integral being defined as the L^2-limit of the Riemann sum. Both T_t and S_t are bounded symmetric operators on $L^2(X;m)$: $\|T_t f\|_{L^2} \leqq \|f\|_{L^2}$, $\|S_t f\|_{L^2} \leqq t\|f\|_{L^2}$. Let $f \in L^1 \cap L^2$. Then $\int_K |T_t f(x)| m(dx) \leqq (T_t I_K, |f|) \leqq \int_X |f(x)| m(dx)$ for any compact set K. Letting $K \uparrow X$, $\|T_t f\|_{L^1} \leqq \|f\|_{L^1}$. Similarly, we have $\|S_t f\|_{L^1} \leqq t\|f\|_{L^1}$. Therefore, T_t and S_t on $L^1 \cap L^2$ can be uniquely extended to bounded linear operators on $L^1(X; m)$ and

$$(1.5.2)\quad \|T_t f\|_{L^1} \leqq \|f\|_{L^1}, \quad \|S_t f\|_{L^1} \leqq t\|f\|_{L^1}, \quad f \in L^1(X; m).$$

The extended operators are positive; moreover, $S_t f(x) \leqq S_{t'} f(x)$ m-a.e. whenever $t < t'$, $f \in L^1_+(X;m)$. Here $L^1_+(X; m) = \{f \in L^1(X; m): f(x) \geqq 0$ m-a.e.$\}$.

Let us put

$$(1.5.3)\quad Gf(x) = \lim_{N\to\infty} S_N f(x) \quad (\leqq +\infty), \quad f \in L^1_+(X; m).$$

The limit exists m-a.e. but may take value $+\infty$. We call the Markovian semigroup $\{T_t, t > 0\}$ *transient* if

$$(1.5.4)\quad Gf(x) < +\infty \quad m\text{-a.e. for any } f \in L^1_+(X; m).$$

Lemma 1.5.1. A Markovian semigroup $\{T_t, t > 0\}$ is transient if and only if there exists a function $g \in L^1(X;m)$ such that g is strictly positive m-a.e. on X and $Gg(x) < +\infty$ m-a.e.

To prove this, we need another lemma:

Lemma 1.5.2 (Hopf's maximal ergodic inequality). For $f \in L^1(X;m)$ and $h > 0$, let $E_h = \{x \in X: \sup_n S_{nh}f(x) > 0\}$. Then $\int_{E_h} S_h f(x) m(dx) \geq 0$.

Proof. Let $E_h^n = \{x \in X: \max_{1 \leq \nu \leq n} S_{\nu h}f(x) > 0\} = \{x \in X: \max_{1 \leq \nu \leq n} (S_{\nu h}f)^+(x) \geq 0\}$. Here we denote $u \vee 0$ by u^+ for a function u on X. Then, for $x \in E_h^n$, we have $S_h f(x) + \max_{1 \leq \nu \leq n} (S_{(\nu+1)h}f - S_h f)^+(x) \geq \max_{1 \leq \nu \leq n} (S_{\nu h}f)^+(x)$. In view of the positivity of T_t, $\max_{1 \leq \nu \leq n} (S_{(\nu+1)h}f - S_h f)^+ \leq T_h [\max_{1 \leq \nu \leq n} (S_{\nu h}f)^+]$. Hence $\int_{E_h^n} S_h f(x) m(dx) \geq \int_{E_h^n} \{\max_{1 \leq \nu \leq n} (S_{\nu h}f)^+(x) - T_h[\max_{1 \leq \nu \leq n} (S_{\nu h}f)^+](x)\} m(dx) \geq \|\max_{1 \leq \nu \leq n} (S_{\nu h}f)^+\|_{L^1} - \|T_h [\max_{1 \leq \nu \leq n} (S_{\nu h}f)^+]\|_{L^1} \geq 0$. Letting n tend to infinity, we have the desired inequality. Q.E.D.

Proof of Lemma 1.5.1. Let g be a function satisfying the condition in Lemma 1.5.1. For any function $f \in L_+^1(X;m)$ and for any $a > 0$ and $h > 0$, we put $A = \{x \in X; \sup_n S_{nh}(f - ag)(x) > 0\}$. Then, by Lemma 1.5.2, $\int_A S_h(f - ag)dm \geq 0$. Since the set $B = \{x \in X; Gf(x) = \infty\}$ is contained in the set A up to an m-negligible set, $h \int_X f dm \geq \int_A S_h f dm \geq a \int_A S_h g dm \geq a \int_B S_h g dm$. Hence $(1/a) \int_X f dm \geq (1/h) \int_{B \cap K} S_h(g \wedge N) dm$ for any compactum K and integer N. Since $(1/h) S_h(g \wedge N)$ is L^2-convergent to $g \wedge N$ as $h \downarrow 0$, $(1/a) \int_X f dm \geq \int_{B \cap K} (g \wedge N) dm$. After letting $K \uparrow X$, $N \uparrow \infty$ and then $a \uparrow \infty$, we get $\int_B g dm = 0$, which implies $m(B) = 0$. Q.E.D.

Let $\{T_t, t > 0\}$ be a strongly continuous Markovian semigroup on $L^2(X;m)$ and $(\mathscr{F}, \mathscr{E})$ be the associated Dirichlet space relative to $L^2(X;m)$.

Lemma 1.5.3. For any non-negative $g \in L^1(X,m) \cap L^2(X;m)$,

$$\sup_{u \in \mathscr{F}} \frac{(|u|, g)}{\sqrt{\mathscr{E}(u, u)}} = \sqrt{\int_X g \cdot Gg dm} \, (\leq +\infty).$$

Proof. First we notice that, for any $g \in L^2(X;m)$, $S_t g \in \mathscr{F}$ and

(1.5.5) $\quad \mathscr{E}(S_t g, u) = (g - T_t g, u), \quad \forall u \in \mathscr{F}.$

Indeed, $S_tg - T_sS_tg = -\int_t^{t+s} T_vgdv + \int_0^s T_vgdv$ and $1/s(S_tg - T_sS_tg, S_tg)$ converges as $s \downarrow 0$ to a finite limit $(g, S_tg) - (T_tg, S_tg)$ proving that $S_tg \in \mathscr{F}$. The same calculation gives (1.5.5).

If $\sup_{u\in\mathscr{F}} (|u|, g)/\sqrt{\mathscr{E}(u, u)} = c < \infty$, then $(S_tg, g) \leqq c\sqrt{\mathscr{E}(S_tg, S_tg)} \leqq c\sqrt{(S_tg, g)}$ and so $\sqrt{(S_tg, g)} \leqq c$. Letting $t \uparrow \infty$, we get $\sqrt{\int_X gGgdm} \leqq c$. Conversely, suppose that $\int_X gGgdm < \infty$. Since $\int_X gGgdm = \int_0^\infty (T_sg, g)\,ds$ and $(T_sg, g) = (T_{s/2}g, T_{s/2}g)$ is non-increasing as $s \uparrow \infty$, we must have $\lim_{s\uparrow\infty} (T_sg, g) = 0$. On account of (1.5.5), we then have $(|u|, g) = \mathscr{E}(|u|, S_tg) + (|u|, T_tg) \leqq \sqrt{\mathscr{E}(S_tg, S_tg)}\sqrt{\mathscr{E}(u, u)} + \sqrt{(T_tg, T_tg)}\sqrt{(u, u)} \leqq \sqrt{(S_tg, g)}\cdot\sqrt{\mathscr{E}(u, u)} + \sqrt{(T_{2t}g, g)}\sqrt{(u, u)} \xrightarrow[t\uparrow\infty]{} \sqrt{\int_X gGgdm}\sqrt{\mathscr{E}(u, u)}$ for any $u \in \mathscr{F}$. Q.E.D.

We say that a Dirichlet space $(\mathscr{F}, \mathscr{E})$ relative to $L^2(X;m)$ is *transient* if there exists a bounded m-integrable function g strictly positive m-a.e. on X such that

$$(1.5.6) \quad \int_X |u| g\, dm \leqq \sqrt{\mathscr{E}(u, u)}, \ \forall u \in \mathscr{F}.$$

The function g above is called a *reference function* of the transient Dirichlet space $(\mathscr{F}, \mathscr{E})$.

Theorem 1.5.1. Let $\{T_t, t > 0\}$ be a strongly continuous Markovian semigroup on $L^2(X;m)$ and $(\mathscr{F}, \mathscr{E})$ be the associated Dirichlet space relative to $L^2(X;m)$. Then $\{T_t, t > 0\}$ is transient if and only if $(\mathscr{F}, \mathscr{E})$ is transient.

Proof. Assume that $\{T_t, t > 0\}$ is transient. Take a strictly positive function $f \in L^1(X;m) \cap C_\infty(X)$ and a sequence of compact sets K_n increasing to X. We then put $\Lambda_n = \{x \in K_n: Gf(x) \leqq n\}$. By assumption $m(X - \bigcup_n \Lambda_n) = 0$. We may assume $m(\Lambda_1) > 0$.

Observe the inequality $\int_{\Lambda_l} GI_{\Lambda_n}dm \leqq (1/c_n)\int_{\Lambda_l} Gfdm \leqq (l/c_n)\, m(\Lambda_l) < \infty$, where $c_n = \inf_{x\in\Lambda_n} f(x)\ (> 0)$. Hence, if we put $\tilde{g}(x) = \sum_{n\geqq 1} \{c_n/nm(\Lambda_n)2^n\} I_{\Lambda_n}(x)$, then $\tilde{g}(x) > 0$ m-a.e., $\tilde{g}(x) \leqq c_1/m(\Lambda_1)$, $\int_X \tilde{g}(x)m(dx) \leqq c_1$ and $\int_X \tilde{g}G\tilde{g}dm = \sum_l \sum_n \dfrac{c_l}{lm(\Lambda_l)2^l}\dfrac{c_n}{nm(\Lambda_n)2^n}\int_{\Lambda_l} GI_{\Lambda_n}dm \leqq$

$$\sum_l \sum_n \frac{c_l}{2^l}\frac{1}{nm(\Lambda_n)2^n} \leqq \frac{c_1}{m(\Lambda_1)} < \infty.$$

In view of Lemma 1.5.3, the Dirichlet space $(\mathscr{F}, \mathscr{E})$ now satisfies (1.5.6) with reference function $g = \tilde{g}/\sqrt{\int_X \tilde{g}\cdot G\tilde{g}dm}$.

Conversely, if $(\mathscr{F},\mathscr{E})$ is transient with a reference function g, then Lemma 1.5.3 implies $\int_X gGg dm \leqq 1$ which means $Gg(x) < \infty$ m-a.e. Therefore, $\{T_t, t > 0\}$ is transient by virtue of Lemma 1.5.1. Q.E.D.

As an immediate consequence of this theorem, we get a simple comparison criterion:

Corollary Let $\{T_t^{(i)}, t > 0\}$, $i = 1, 2$, be strongly continuous Markovian semigroups on $L^2(X;m)$ and suppose that the associated Dirichlet spaces satisfy $\mathscr{F}^{(1)} \supset \mathscr{F}^{(2)}$, $\mathscr{E}^{(1)}(u, u) \leqq C \cdot \mathscr{E}^{(2)}(u,u)$, $u \in \mathscr{F}^{(2)}$, C being some positive constant. If $\{T_t^{(1)}, t>0\}$ is transient, then so is $\{T_t^{(2)}, t>0\}$.

Theorem 1.5.1 tempts us into introducing the following notion. A pair $(\mathscr{F}_e,\mathscr{E})$ is called an *extended (transient) Dirichlet space with reference measure m* if the following conditions are satisfied:

($\mathscr{F}_e$.1) $\mathscr{F}_e$ is a real Hilbert space with inner product $\mathscr{E}$,

($\mathscr{F}_e$.2) there exists an m-integrable bounded function g strictly positive m-a.e. such that $\mathscr{F}_e \subset L^1(X;g \cdot m)$ and $\int_X |u(x)| g(x) m(dx) \leqq \sqrt{\mathscr{E}(u, u)}$, $\forall u \in \mathscr{F}_e$,

($\mathscr{F}_e$.3) $\mathscr{F}_e \cap L^2(X;m)$ is dense both in $L^2(X;m)$ and in $(\mathscr{F}_e, \mathscr{E})$,

($\mathscr{F}_e$.4) every normal contraction operates on $(\mathscr{F}_e, \mathscr{E})$; i.e., $u \in \mathscr{F}_e$, v is a normal contraction of $u \Rightarrow v \in \mathscr{F}_e$, $\mathscr{E}(v, v) \leqq \mathscr{E}(u,u)$.

The function g in ($\mathscr{F}_e$.2) is called a *reference function* of the extended Dirichlet space $(\mathscr{F}_e,\mathscr{E})$.

Theorem 1.5.2. (i) Assume that $(\mathscr{F}_e,\mathscr{E})$ is an extended Dirichlet space with reference measure m. Put $\mathscr{F} = \mathscr{F}_e \cap L^2(X;m)$. Then $(\mathscr{F},\mathscr{E})$ is a transient Dirichlet space relative to $L^2(X;m)$. $\mathscr{F}_e$ is the completion of $\mathscr{F}$ with respect to $\mathscr{E}$.

(ii) Conversely, assume $(\mathscr{F},\mathscr{E})$ is a transient Dirichlet space relative to $L^2(X;m)$. Then the completion $(\mathscr{F}_e,\mathscr{E})$ of $(\mathscr{F},\mathscr{E})$ is an extended Dirichlet space with reference measure m. Moreover, $\mathscr{F} = \mathscr{F}_e \cap L^2(X;m)$.

The first assertion of this theorem is clear from the definition. The second assertion is amplified in the next lemma.

Lemma 1.5.4. Let $(\mathscr{F},\mathscr{E})$ be a transient Dirichlet space on $L^2(X;m)$ with reference function g. Let $(\mathscr{F}_e,\mathscr{E})$ be the abstract completion of $(\mathscr{F},\mathscr{E})$.

(i) $\mathscr{F}_e$ is embedded in $L^1(X;g \cdot m)$ and the inequality (1.5.6) holds for any $u \in \mathscr{F}_e$.

(ii) Extend the expression (1.4.8) of the approximating form associated with $(\mathcal{F},\mathcal{E})$ to any Borel function u on X:

$$\mathcal{E}^{(\beta)}(u, u) = \frac{1}{2}\beta \int_X \int_X (u(x) - u(y))^2 \sigma_\beta(dx, dy) + \beta \int_X u(x)^2 (1 - s_\beta(x)) m(dx).$$

If $u \in \mathcal{F}_e$, then for any Borel modification $\tilde{u}$ of u, $\mathcal{E}^{(\beta)}(\tilde{u}, \tilde{u})$ increases to $\mathcal{E}(u, u)$ as $\beta \uparrow \infty$.

(iii) Every normal contraction operates on $(\mathcal{F}_e,\mathcal{E})$.

(iv) $\mathcal{F} = \mathcal{F}_e \cap L^2(X;m)$.

Proof. Let $u \in \mathcal{F}_e$ represent an $\mathcal{E}$-Cauchy sequence $u_n \in \mathcal{F}$. In view of inequality (1.5.6), u_n is $L^1(X;g\cdot m)$-convergent to some element $\theta u, \in L^1(X;g\cdot m)$. We may assume that some Borel modifications $\tilde{u}_n$ and $\widetilde{\theta u}$ satisfy $\lim_{n\to\infty} \tilde{u}_n(x) = \widetilde{\theta u}(x)$, $\forall x \in X$. Then by Fatou's lemma $\mathcal{E}^{(\beta)}(\tilde{u}_n - \widetilde{\theta u}\ \tilde{u}_n - \widetilde{\theta u}) \leq \underline{\lim}_{m\to\infty} \mathcal{E}^{(\beta)}(\tilde{u}_n - \tilde{u}_m, \tilde{u}_n - \tilde{u}_m) \leq \underline{\lim}_{m\to\infty} \mathcal{E}(u_n - u_m, u_n - u_m)$ which can be made arbitrarily small for sufficiently large n. In particular, $\lim_{n\to\infty} \mathcal{E}^{(\beta)}(\tilde{u}_n, \tilde{u}_n) = \mathcal{E}^{(\beta)}(\widetilde{\theta u}, \widetilde{\theta u}) < \infty$ and, hence, $\mathcal{E}^{(\beta)}(\widetilde{\theta u}, \widetilde{\theta u})$ is non-decreasing with β. Furthermore, the inequality $|\sqrt{\mathcal{E}(u, u)} - \sqrt{\mathcal{E}^{(\beta)}(\widetilde{\theta u}, \widetilde{\theta u})}| \leq |\sqrt{\mathcal{E}(u, u)} - \sqrt{\mathcal{E}(u_n, u_n)}| + |\sqrt{\mathcal{E}(u_n, u_n)} - \sqrt{\mathcal{E}^{(\beta)}(\tilde{u}_n, \tilde{u}_n)}| + \sqrt{\mathcal{E}^{(\beta)}(\tilde{u}_n - \widetilde{\theta u}, \tilde{u}_n - \widetilde{\theta u})}$ implies that $\lim_{\beta\uparrow\infty} \mathcal{E}^{(\beta)}(\widetilde{\theta u}, \widetilde{\theta u}) = \mathcal{E}(u, u)$. This means that $u = 0$ if $\theta u = 0$ m-a.e. Thus we can identify $u \in \mathcal{F}_e$ with $\theta u \in L^1(X;g\cdot m)$ and both statements (i) and (ii) are proven.

To verify (iii), suppose that v is a normal contraction of $u \in \mathcal{F}_e$. Choose $u_n \in \mathcal{F}$ which is $\mathcal{E}$-convergent to u. Set $w_n = (v \wedge |u_n|)^+$. Then it is easy to see that $\mathcal{E}^{(\beta)}(w_n, w_n) \leq \mathcal{E}^{(\beta)}(\tilde{v}, \tilde{v}) + \mathcal{E}^{(\beta)}(u_n, u_n) \leq \mathcal{E}(u, u) + \mathcal{E}(u_n, u_n)$. Since $w_n \in L^2(X;m)$, w_n belongs to $\mathcal{F}$ and $\mathcal{E}(w_n, w_n)$ is uniformly bounded in n. Therefore, the Cesàro mean of a subsequence of w_n is $\mathcal{E}$-convergent to some $w \in \mathcal{F}_e$.[1)] But (i) implies $v^+ = w \in \mathcal{F}_e$ and so $v = v^+ - v^- \in \mathcal{F}_e$.

(iv) is an immediate consequence of (ii). Q.E.D.

Theorem 1.5.3. The family of all extended Dirichlet spaces $(\mathcal{F}_e, \mathcal{E})$ with reference measure m and the family of all strongly continuous transient Markovian semigroups $\{T_t, t > 0\}$ on $L^2(X;m)$ are in one to one correspondence. The correspondence is characterized by the following relation: for any $f \in L^1_+(X;m) \cup L^2_+(X;m)$ such that $\int_X f\cdot Gf dm < \infty$,

(1.5.7) $Gf \in \mathcal{F}_e$, $\mathcal{E}(Gf, v) = (f, v)$, $\forall v \in \mathcal{F}_e$.

---

1) Cf. F. Riesz and B. Sz. Nagy [76; §38].

Proof. Let a transient semigroup $\{T_t, t>0\}$ and an extended Dirichlet space $(\mathscr{F}_e, \mathscr{E})$ correspond to each other as in Theorem 1.5.1 and Theorem 1.5.2. First, consider a non-negative $f \in L^2(X;m)$ such that $\int fGfdm < \infty$. By (1.5.5) we have for $t > t'$
$\mathscr{E}(S_t f - S_{t'} f,\ S_t f - S_{t'} f) \leq (S_t f,\ f) - (S_{t'} f, f) + (S_{t'} f,\ T_t f - T_{t'} f)$ which converges as $t,\ t' \to \infty$ to zero because $(S_t f, f) \to \int fGfdm$ and $(S_{t'}, f, T_t f) = \int_t^{t+t'} (T_u f, f)du \to 0$. Hence $S_t f$ is $\mathscr{E}$-convergent in $\mathscr{F}_e$ and the limit coincides with Gf by virtue of $(\mathscr{F}_e.2)$. Now (1.5.5) again leads us to the equation (1.5.7).

Next consider $f \in L^1_+$ such that $\int fGfdm < \infty$ and put $f_n = f \wedge n$. Applying (1.5.7) to $f_n \in L^2(X;m)$, $\mathscr{E}(Gf_n - Gf_m, Gf_n - Gf_m) \leq \int f_n Gf_n dm - \int f_m Gf_m dm$, $n > m$, which converges to zero as $n, m \to \infty$. We can then repeat a similar argument as above to get (1.5.7).

For a given transient semigroup $\{T_t,\ t > 0\}$, the extended Dirichlet space $(\mathscr{F}_e, \mathscr{E})$ satisfying (1.5.7) is unique because functions Gf taking part in (1.5.7) are dense in $(\mathscr{F}_e, \mathscr{E})$.

Conversely, given $(\mathscr{F}_e, \mathscr{E})$, the transient semigroup $\{T_t, t > 0\}$ satisfying (1.5.7) is unique. To see this, observe that $Gf = \lim_{n\to\infty} G_{1/n} f$ m-a.e. for non-negative $f \in L^2 \cap L^1$ and so the resolvent equation extend to $Gf = G_\alpha f + \alpha G G_\alpha f$. Assume further that $\int fGfdm < \infty$. Then $\int G_\alpha f \cdot GG_\alpha f dm \leq (1/\alpha) \int G_\alpha f \cdot Gf dm = (1/\alpha) \int fGG_\alpha f dm \leq (1/\alpha^2) \int fGfdm < \infty$. In view of (1.5.7), $GG_\alpha f \in \mathscr{F}_e$ and $\mathscr{E}(GG_\alpha f,\ v) = (G_\alpha f, v)$, $\forall v \in \mathscr{F}_e$. Hence $(f, v) = \mathscr{E}(Gf,\ v) = \mathscr{E}(G_\alpha f,\ v) + \alpha \mathscr{E}(GG_\alpha f,\ v) = \mathscr{E}(G_\alpha f,\ v) + \alpha(G_\alpha f,\ v)$, namely, $\mathscr{E}_\alpha(G_\alpha f,\ v) = (f,\ v)$, $\forall v \in \mathscr{F}$, which means that $\{G_\alpha,\ \alpha > 0\}$ is uniquely determined by $(\mathscr{F}_e,\ \mathscr{E})$ and so is $\{T_t, t > 0\}$. Q.E.D.

A. Beurling and J. L. Deny[1)] first introduced the notion of Dirichlet space. In the present context Beurling-Deny's original notion is nothing but an extended Dirichlet space $(\mathscr{F}_e, \mathscr{E})$ which satisfies instead of $(\mathscr{F}_e.2)$ and $(\mathscr{F}_e.3)$ the stronger conditions $(\mathscr{F}_e.2)'$ and $(\mathscr{F}_e.3)'$ described below:

$(\mathscr{F}_e.2)'$ $\mathscr{F}_e \subset L^1_{loc}(X;m)$ and for any compact set $K \subset X$

$$\int_K |u(x)|\, m(dx) \leq C_K \sqrt{\mathscr{E}(u, u)},\ \forall u \in \mathscr{F}_e,$$

C_K being a positive constant.

$(\mathscr{F}_e.3)'$ $(\mathscr{F}_e, \mathscr{E})$ is regular; i.e., $\mathscr{F}_e \cap C_0(X)$ is dense both in $(\mathscr{F}_e, \mathscr{E})$ and in $C_0(X)$.

Notice that the condition $(\mathscr{F}_e.2)'$ is the same as saying that $(\mathscr{F}_e, \mathscr{E})$

1) A. Beurling and J. L. Deny [12].

possesses as its reference function g a strictly positive function of $C_\infty(X) \cap L^1(X;m)$. The next theorem is an immediate consequence of Lemma 1.5.3 and Lemma 1.5.4(i).

Theorem 1.5.4. Let an extended Dirichlet space $(\mathscr{F}_e, \mathscr{E})$ with reference measure m and a strongly continuous transient Markovian semigroup $\{T_t, t > 0\}$ correspond to each other as in Theorem 1.5.3. Then a necessary and sufficient condition that $(\mathscr{F}_e, \mathscr{E})$ possesses the stronger property $(\mathscr{F}_e. 2)'$ is that

$$\int f \cdot Gf\, dm \left(= \int_0^\infty (T_t f, f) dt \right) < \infty, \quad \forall f \in C_0^+(X).^{1)}$$

Example 1.5.1. Consider Example 1.2.3. and assume that the Euclidean domain D is bounded. Then Poincaré's inequality[2]

$$(1.5.8) \quad (u, u) \leqq \frac{d^2}{2n} \mathbf{D}(u, u), \qquad u \in H_0^1(D)$$

holds with d being the diameter of D. Hence the Dirichlet space $(\mathscr{F}, \mathscr{E})$ with $\mathscr{F} = H_0^1(D)$, $\mathscr{E} = 1/2\mathbf{D}$, is transient. Since $\mathscr{F}$ is already complete with respect to $\mathscr{E}$, $\mathscr{F}_e = \mathscr{F} = H_0^1(D)$. G is bounded on $L^2(D)$.

Example 1.5.2 Consider a symmetric convolution semigroup $\{\nu_t, t > 0\}$ on R^n as in Example 1.4.1. The associated semigroup $\{T_t, t > 0\}$ and Dirichlet space $(\mathscr{F}, \mathscr{E})$ are defined by (1.4.20) and (1.4.24) respectively. We put $\kappa(A) = \int_0^\infty \nu_t(A) dt (\leqq +\infty)$ and assert that the following four conditions are equivalent:

(1.5.9) $\{T_t, t > 0\}$ is transient.

(1.5.10) $\kappa(A) < \infty$ for any compact set $A \subset R^n$.

(1.5.11) $\int_0^\infty (T_t f, f) dt < \infty, \quad \forall f \in C_0^+(R^n)$.[3]

(1.5.12) $1/\psi(\xi)$ is locally integrable on R^n.

Assume (1.5.9). Then $\int_0^\infty p_t(x, A) dt = \kappa(A - x)$ is finite for any compact A and for almost all x. This means (1.5.10), which in turn implies (1.5.11).

1) $C_0^+(X) = \{u \in C_0(X) \colon u \geqq 0\}$.
2) Cf. S. Mizohata [69; pp. 169].
3) $C_0^+(R^n)$ is the space of all non-negative continuous functions on R^n with compact support.

Obviously, (1.5.9) follows from (1.5.11). Let A be any compact set and $f \in C_0^+(R^n)$ be such that $|\hat{f}(x)| \geqq 1$, $\forall x \in A$. Then $\int_A \frac{d\xi}{\psi(\xi)} \leqq \int_{R^n} \frac{|\hat{f}(\xi)|^2}{\psi(\xi)} d\xi = \int_0^\infty \widehat{(T_t f, \hat{f})}\, dt = \int_0^\infty (T_t f, f) dt$, which is finite under the condition (1.5.11). On the other hand, by putting

$$g(\xi) = C\int_{R^n} f(\xi - \eta) f(\eta) d\eta, \text{ with } C = (2\pi)^{-n/2},$$

$$\kappa(A) \leq \lim_{T\to\infty}\int_0^T \left\{\int_{R^n} |\hat{f}(x)|^2 \nu_t(dx)\right\} dt = \lim_{T\to\infty}\int_0^T \left\{\int_{R^n} \check{g}(x)\nu_t(dx)\right\} dt =$$

$$C \lim_{T\to\infty}\int_0^T \left\{\int_{R^n} g(\xi)\hat{\nu}_t(\xi) d\xi\right\} dt = C\int_{R^n} \frac{g(\xi)}{\psi(\xi)} d\xi.$$

In view of Theorem 1.5.4 and (1.5.11), the extended Dirichlet space $(\mathscr{F}_e, \mathscr{E})$ associated with a transient convolution semigroup always satisfies the Beurling-Deny condition $(\mathscr{F}_e.2)'$. It satisfies the regularity condition $(\mathscr{F}_e.3)'$ as well because $(\mathscr{F}, \mathscr{E})$ is regular as we saw in Example 1.4.1. Furthermore, the following explicit expression is known:

$$(1.5.13)\quad \begin{cases} \mathscr{F}_e = \{u \in L^1_{loc}(R^n): u \text{ is a tempered distribution and } \\ \qquad \hat{u} \in L^2(\psi \cdot d\xi)\} \\ \mathscr{E}(u, v) = \int_{R^n} \hat{u}(\xi)\bar{\hat{v}}(\xi)\psi(\xi) d\xi, \quad u, v \in \mathscr{F}_e. \end{cases}$$

Since $(\mathscr{F}_e, \mathscr{E})$ is the completion of $(\mathscr{F}, \mathscr{E})$ of (1.4.24) on account of Theorem 1.5.2, $(\mathscr{F}_e \mathscr{E})$ is isometric to the space $L^2(\psi \cdot d\xi)$. On the other hand, we can see that any element of $L^2(\psi \cdot d\xi)$ is slowly increasing and that the convergence in $L^2(\psi \cdot d\xi)$ implies the convergence as tempered distributions. This follows easily from the observation that $1/\psi(\xi)$ is slowly increasing.[1] We thus get (1.5.13).

The semigroup corresponding to $\psi(\xi) = c|\xi|^\alpha$, $0 < \alpha \leqq 2$, $c > 0$, is called a *symmetric stable semigroup* of index α. According to (1.5.12), a symmetric stable semigroup is transient if and only if $\alpha < n$. In this case $(\mathscr{F}_e, \mathscr{E})$ can be described as follows:

$$(1.5.14)\quad \begin{cases} \mathscr{F}_e = \{u = R^{(\alpha/2)} f : f \in L^2(R^n)\} \\ \mathscr{E}(u, v) = \int_{R^n} f(x) g(x)\, dx \quad \text{for} \quad u = R^{(\alpha/2)} f, \quad v = R^{(\alpha/2)} g, \end{cases}$$

where $R^{(\alpha)}$ denotes the Riesz kernel of index α:

$$(1.5.15)\quad R^{(\alpha)}(x) = \frac{\Gamma((n-\alpha)/2)}{c 2^\alpha \pi^{n/2} \Gamma(\alpha/2)} |x|^{\alpha - n}.$$

$\mathscr{F}_e$ is namely the set of all Riesz potentials (of index $\alpha/2$) of L^2-functions.

In order to verify (1.5.14), we note that 0-order Green measure $\kappa(A) = \int_0^\infty \nu_t(A) dt$ is given by

---

1) Cf. J. Deny [25; pp. 194].

$$(1.5.16)\quad \kappa(A) = \int_A R^{(\alpha)}(x)dx$$

because both measures possess the same Fourier transform $1/c|\xi|^{\alpha}$. By virtue of Theorem 1.5.3 and (1.5.11), $(\mathscr{F}_e,\mathscr{E})$ is characterized as a real Hilbert space for which $R^{(\alpha)}f \in \mathscr{F}_e$, $\mathscr{E}(R^{(\alpha)}f, v) = \int_{R^n} f(x)v(x)dx$, $\forall f \in C_0(R^n)$, $\forall v \in \mathscr{F}_e$. It is easy to see that the space defined by the right hand side of (1.5.14) has this property because $R^{(\alpha)}f$ can be written as $R^{(\alpha/2)}g$ with $g = R^{(\alpha/2)}f \in L^2(R^n)$ for any $f \in C_0(R^n)$. Thus, (1.5.14) is proven.

As an application of (1.5.14), we can rewrite the Sobolev inequality[1)] $\|R^{(\alpha/2)}f\|_{p_0} \leq C\cdot\|f\|_2$, $f \in L^2(R^n)$, $1/p_0 = 1/2 - \alpha/2n$, to get

$$(1.5.17)\quad \|u\|_{p_0} \leq C\sqrt{\mathscr{E}(u, u)}, \qquad u \in \mathscr{F}_e,$$

where C is a positive constant and $\|\ \|_p$ denotes L^p-norm. (1.5.17) is stronger than our basic inequality $(\mathscr{F}_e.2)'$ for the space $(\mathscr{F}_e,\mathscr{E})$. In fact, it means that the space $(\mathscr{F}_e, \mathscr{E})$ is always embedded continuously into $L^2_{loc}(R^n)$.

The Dirichlet space $(1/2\,\mathbf{D}, H^1(R^n))$ corresponds to $\alpha = 2$ and $c = 1/2$. When $n \geqq 3$, this is transient and the corresponding extended Dirichlet space $(\mathscr{F}_e,\mathscr{E})$ is the completion of $C_0^\infty(R^n)$ with the Dirichlet integral $1/2\,\mathbf{D}$. Owing to its property $(\mathscr{F}_e.2)'$, we can see that $\mathscr{F}_e$ is a closed subspace of the function space

$$(1.5.18)\quad \mathscr{G} = \{u \in L^1_{loc}(R^n) : \frac{\partial u}{\partial x_i} \in L^2(R^n), 1 \leq i \leq n\}$$

with $\mathscr{E}(u, v) = 1/2\,\mathbf{D}(u, v)$, $u, v \in \mathscr{F}_e$, the derivatives being taken in the distribution sense.

The associated 0-order potential is given by the Newtonian kernel

$$(1.5.19)\quad \boldsymbol{n}(x) = \frac{\Gamma(n/2 - 1)}{2\pi^{n/2}}|x|^{2-n}.$$

The Newtonian potential $Nf(x) = \int_{R^n} \boldsymbol{n}(x - y)f(y)dy$ for $f \in C_0(R^n)$ belongs to $\mathscr{F}_e$ and $1/2\,\mathbf{D}(Nf, v) = (f, v)$, $\forall v \in \mathscr{F}_e$. This property characterizes the Hilbert space $(\mathscr{F}_e, \mathscr{E})$.

1) Cf. E. M. Stein [82; Chap. V].

CHAPTER 2

The Scope of Dirichlet Forms

The Markovian nature and the local property of a closable symmetric form are preserved under the operation of taking the smallest closed extension (§ 2.1).

The integro-differential form exhibited in §1.2 is the most general expression of the closable Markovian symmetric form (§2.2).

In general, a closable Markovian symmetric form admits many closed extensions larger than the smallest one. They are not necessarily Markovian, but there is a maximum Markovian extension among them (§2.3).

Those examples of closable symmetric forms exhibited in §2.1 shall give rise to diffusion processes in §6.2. The probabilistic meanings of those measures appearing in the Beurling-Deny first formula (2.2.1) shall be specified in §4.5. The Beurling-Deny second formula (2.2.13) shall be derived again in §5.4 by using a stochastic calculus.

§2.1 Closability and the Smallest Closed Extension

Let X and m be as in § 1.1. We show that the Markovian nature and the local property of the form are preserved under the operation of taking the smallest closed extension.

Theorem 2.1.1. Suppose $\mathscr{E}$ is a closable Markovian symmetric form on $L^2(X;m)$. Then its smallest closed extension $\bar{\mathscr{E}}$ is again Markovian and hence a Dirichlet form.

Proof. Let $\{\alpha G_\alpha, \alpha > 0\}$ be the strongly continuous resolvent associated with the closed form $\bar{\mathscr{E}}$. By virtue of Theorem 1.4.1, it suffices to show that αG_α is Markovian.

Take $u \in L^2(X;m)$ such that $0 \leq u \leq 1$ m-a.e. and consider the quadratic functional ψ defined by (1.4.1). By (1.4.2), we can see that $G_\alpha u$ is the unique element minimizing ψ on $\mathscr{D}[\bar{\mathscr{E}}]$ and that $v_n \in \mathscr{D}[\bar{\mathscr{E}}]$ is $\bar{\mathscr{E}}_1$-convergent to $G_\alpha u$ if and only if $\lim_{n\to\infty} \psi(v_n) = \psi(G_\alpha u)$.

We can select such a sequence $\{v_n\}$ from $\mathscr{D}[\mathscr{E}]$ because $\mathscr{D}[\mathscr{E}]$ is $\mathscr{E}_1$-dense in $\mathscr{D}[\bar{\mathscr{E}}]$. For any $\varepsilon > 0$, let $\phi_\varepsilon(t)$ be a real function as given in the Markovian condition ($\mathscr{E}$.4) in §1.1 and put $\tilde{\phi}_\varepsilon(t) = (1/\alpha)\,\phi_{\alpha\varepsilon}(\alpha t)$ and $w_n = \tilde{\phi}_\varepsilon(v_n)$. In the same way as in the proof of the implication (c) $\Rightarrow$ (b) of Theorem 1.4.1, we have $\psi(w_n) \leqq \psi(v_n)$. Therefore, $\lim_{n\to\infty} \psi(w_n) = \psi(G_\alpha u)$ and w_n is $\mathscr{E}_1$-convergent to $G_\alpha u$. In particular, a subsequence of w_n converges to $G_\alpha u$ m-a.e. Since $-\varepsilon \leqq w_n \leqq 1/\alpha + \varepsilon$ m-a.e., $-\varepsilon \leqq G_\alpha u \leqq 1/\alpha + \varepsilon$ m-a.e., for arbitrary ε, we get the Markovian nature of αG_α. Q.E.D.

Theorem 2.1.2 Assume that a closable Markovian symmetric form on $L^2(X;m)$ satisfies the following conditions (i) and (ii):
(i) $\mathscr{D}[\mathscr{E}]$ is a dense subalgebra of $C_0(X)$,
(ii) there exists for any compact set K and relatively compact open set $G \supset K$ a non-negative function $u \in \mathscr{D}[\mathscr{E}]$ such that $u(x) = 1$ for $x \in K$ and $u(x) = 0$ for $x \in X - G$.

If $\mathscr{E}$ has the local property, then so does its smallest closed extension $\bar{\mathscr{E}}$. Furthermore, $\bar{\mathscr{E}}$ is a regular Dirichlet form.

Proof. The last statement is clear from Theorem 2.1.1 and condition (i). Assuming the local property of $\mathscr{E}$, we claim the same property of $\bar{\mathscr{E}}$:

(2.1.1) $\bar{\mathscr{E}}(u, v) = 0$ if $u, v \in \mathscr{D}[\bar{\mathscr{E}}]$ have disjoint compact supports.

We only give the proof of (2.1.1) when $v \in \mathscr{D}[\mathscr{E}]$ because the general case that $v \in \mathscr{D}[\bar{\mathscr{E}}]$ can be proved in a similar way. We may further assume $0 \leqq u \leqq 1$ m-a.e. In fact, by putting $u_l^+ = (0 \vee u) \wedge l$, $u_l^- = ((-l) \vee u) \wedge 0$ for a general $u \in \mathscr{D}[\bar{\mathscr{E}}]$, we have $\mathrm{Supp}[u_l^+] \subset \mathrm{Supp}[u]$, $\mathrm{Supp}[u_l^-] \subset \mathrm{Supp}\,[u]$, and $\lim_{l\to\infty} \{\bar{\mathscr{E}}(u_l^+, v) + \bar{\mathscr{E}}(u_l^-, v)\} = \bar{\mathscr{E}}(u, v)$ by Theorem 1.4.2.

Let us prove (2.1.1) under the above assumptions. Choose $u_n \in \mathscr{D}[\mathscr{E}]$ $\mathscr{E}_1$-convergent to u. By virtue of (ii), there exists $w \in \mathscr{D}[\mathscr{E}]$ such that $w(x) = 1$, $\forall x \in \mathrm{Supp}[u]$ and $\mathrm{Supp}[w] \cap \mathrm{Supp}[v] = \phi$. Put $\tilde{u}_n = \phi_{1/n}(u_n)\cdot w$ with $\phi_{1/n}$ being the function in the Markovian condition ($\mathscr{E}$.4). Then $\tilde{u}_n \in \mathscr{D}[\mathscr{E}]$ is L^2-convergent to u. Furthermore $\mathscr{E}_1(\tilde{u}_n, \tilde{u}_n)$ is uniformly bounded because $\sqrt{\mathscr{E}_1(\tilde{u}_n, \tilde{u}_n)} \leqq (1 + 1/n)\sqrt{\mathscr{E}_1(w, w)} + \|w\|_\infty \sqrt{\mathscr{E}_1(u_n; u_n)}$ by Theorem 1.4.2. Therefore, $\tilde{u}_n$ is weakly convergent to u with respect to $\bar{\mathscr{E}}_1$; in particular, $\bar{\mathscr{E}}_1(u, v) = \lim_{n\to\infty} \mathscr{E}_1(\tilde{u}_n, v) = 0$. Q.E.D.

In the remainder of this section, we pay special attention to the Markovian symmetric form

$$(2.1.2)\quad \begin{cases} \mathscr{E}(u, v) = \sum_{i,j=1}^{n} \int_D \frac{\partial u(x)}{\partial x_i} \frac{\partial v(x)}{\partial x_j} \nu_{ij}(dx) \\ \mathscr{D}[\mathscr{E}] = C_0^\infty(D) \end{cases}$$

which appeared in Example 1.2.1 and ask ourselves the question "Under what conditions on ν_{ij} is the form (2.1.2) closable on $L^2(D)$?" When (2.1.2) is closable, then its smallest closed extension gives rise to a regular Dirichlet form on $L^2(D)$ with the local property[1)] on account of the preceding two theorems.

In the one dimensional case a satisfactory answer to the above question is known. We start however with the multidimensional cases.

(1°) Absolutely continuous case.

Consider the case where ν_{ij} is absolutely continuous, i.e.,

$$(2.1.3)\quad \nu_{ij}(dx) = a_{ij}(x)dx, \qquad 1 \leqq i, \ j \leqq n,$$

where a_{ij} are locally integrable Borel functions on D such that for any $\xi \in R^n$ and $x \in D$

$$(2.1.4)\quad \sum_{i,j=1}^{n} a_{ij}(x)\xi_i\xi_j \geqq 0, \quad a_{ij}(x) = a_{ji}(x), \quad 1 \leqq i, j \leqq n.$$

If $\{a_{ij}\}$ satisfies one of the following two conditions, then the form (2.1.2) is closable on $L^2(D)$:

(1°. a) $\frac{\partial}{\partial x_k} a_{ij} \in L^1_{loc}(D), \quad 1 \leqq i, j, k \leqq n.$

(1°. b) uniform ellipticity: $\exists\delta > 0, \quad \sum_{i,j=1}^{n} a_{ij}(x)\xi_i\xi_j \geqq \delta|\xi|^2, \ \forall \xi \in R^n, \forall x \in D.$

Here the derivatives are taken in the distribution sense.

Assume **(1°.a)**. Then $Su(x) = \sum_{i,j=1}^{n} \frac{\partial}{\partial x_i}\left(a_{ij}(x)\frac{\partial u(x)}{\partial x_j}\right)$, $\mathscr{D}(S) = C_0^\infty(D)$, defines a symmetric operator S on $L^2(D)$ and $\mathscr{E}(u, v) = -(u, Sv)$, $u, v \in \mathscr{D}[\mathscr{E}]$. This expression implies that $\mathscr{E}$ satisfies the condition (1.1.3). Hence $\mathscr{E}$ is closable.

Assume **(1°. b)** instead of **(1°. a)**. Suppose $u_l \in C_0^\infty(D)$ satisfies $\lim_{l,m\to\infty} \mathscr{E}(u_l - u_m, u_l - u_m) = 0$ and $\lim_{l\to\infty}(u_l, u_l) = 0$; then $\{u_l\}$ constitutes a Cauchy sequence with respect to the Dirichlet integral **D**. Since **D** is a special

1) Cf. Theorem 6.2.2.

form satisfying condition (**1°. a**), we have $\lim_{l\to\infty} \mathbf{D}(u_l, u_l) = 0$. Subtracting a subsequence if necessary, we may assume $\partial u_l/\partial x_i$ converges to zero a.e. on D, $1 \leqq i \leqq n$. Then by Fatou's lemma $\mathscr{E}(u_m, u_m) = \int_D \lim_{l\to\infty} \left(\sum_{i,j=1}^n \frac{\partial(u_l - u_m)}{\partial x_i} \frac{\partial(u_l - u_m)}{\partial x_j} a_{ij} \right) dx \leqq \varliminf_{l\to\infty} \mathscr{E}(u_l - u_m, u_l - u_m)$, which can be made arbitrarily small for a sufficiently large m.

Here we give another closable case in which a_{ij} may satisfy neither (**1°. a**) nor (**1°. b**). Assume $D = R^n$ and assume further that for any $1 \leqq i, j \leqq n$

(**1°. c**) the distribution derivative $(\partial/\partial x_k)\, a_{ij}$ belongs to $L^1_{loc}(R^n)$ for $1 \leqq k \leqq n-1$, but the derivative $(\partial/\partial x_n)\, a_{ij}(x)$ exists in the ordinary sense only on $\{x \in R^n;\, x_n \neq 0\}$.

Then (2.1.2) is closable. To see this, put $x' = (x_1, \ldots, x_{n-1})$ for $x = (x_1, \ldots, x_{n-1}, x_n)$ and set
$\mathscr{F}^0 = \{u \in C_0^\infty(R^n);\, {}^\exists\delta > 0, u(x) \text{ depends only on } x' \text{ when } x_n \in (-\delta, \delta)\}$.
In the same way as in the proof (**1°.a**), we can see that the form (2.1.2) with the domain $\mathscr{D}[\mathscr{E}]$ being restricted to $\mathscr{F}^0$ is closable on $L^2(R^n)$. Next define for any $u \in C_0^\infty(R^n)$

$$v_\delta(x) = \begin{cases} u(x', 0) & -2\delta \leqq x_n \leqq 2\delta \\ u(x', x_n - 2\delta) & x_n > 2\delta \\ u(x', x_n + 2\delta) & x_n < -2\delta \end{cases}$$

and put $u_\delta(x) = j_\delta * v_\delta(x)$ with the mollifier j_δ of Problem 1.2.1. Obviously $u_\delta \in \mathscr{F}^0$ and Supp $[u_\delta]$ is contained in a compact set independent of $\delta > 0$. When $\delta_n \downarrow 0$, $u_{\delta_n}(x) \to u(x)$ and $(\partial/\partial x_k)\, u_{\delta_n}(x) \to \partial u(x)/\partial x_k$ boundedly on the set $\{x \in R^n;\, x_n \neq 0\}$. Hence $\lim_{n\to\infty} \mathscr{E}_1(u_{\delta_n} - u, u_{\delta_n} - u) = 0$, proving the closability of (2.1.2).

The following example is a special case of (2.1.2) satisfying (**1°. C**):

$$(2.1.5)\quad \begin{cases} \mathscr{E}(u, v) = C_1 \int_{R_+^n} \sum_{i=1}^n \frac{\partial u}{\partial x_i} \frac{\partial v}{\partial x_i} dx + C_2 \int_{R_-^n} \sum_{i=1}^n \frac{\partial u}{\partial x_i} \frac{\partial v}{\partial x_i} dx \\ \mathscr{D}[\mathscr{E}] = C_0^\infty(R^n) \end{cases}$$

where $C_1, C_2 \geqq 0$, $R_+^n = \{x \in R^n;\, x_n > 0\}$ and $R_-^n = \{x \in R^n;\, x_n < 0\}$.

(2°) Superposition of closable forms.

In order to treat the cases where the ν_{ij} are not necessarily absolutely continuous with respect to the Lebesgue measure, we formulate a method of superposing closable symmetric forms to get new closable symmetric forms inductively.

Assume $D = R^n$ for simplicity and let $\mathscr{E}$ be a symmetric form on $L^2(R^n)$ coercive on the Sobolev space of Example 1.2.3 and specifically expressed as follows:

$$(2.1.6)\quad \begin{cases} \mathscr{E}(u, v) = \int_\Theta \mathscr{E}^\theta(u^\theta, v^\theta)\mu(d\theta) + c\cdot\mathbf{D}(u, v) \\ \mathscr{D}[\mathscr{E}] = C_0^\infty(R^n). \end{cases}$$

c being a positive constant. Here we are given an auxiliary σ-finite measure space $(\Theta, \mathscr{B}(\Theta), \mu)$, a collection $\{F_\theta;\ \theta \in \Theta\}$ of $(n-1)$-dimensional hyperplanes F_θ, and, for each $\theta \in \Theta$, a symmetric form $\mathscr{E}^\theta$ on $L^2(F_\theta)$ with $\mathscr{D}[\mathscr{E}^\theta] = C_0^\infty(F_\theta)$, $L^2(F_\theta)$ being based on the $(n-1)$-dimensional Lebesgue measure on F_θ. We assume that $\mathscr{E}^\theta(u^\theta, u^\theta)$ is μ-integrable in $\theta \in \Theta$ for each $u \in C_0^\infty(R^n)$, where u^θ is the restriction to F_θ of u.

Theorem 2.1.3 (i) Assume $\mathscr{E}^\theta$ is closable on $L^2(F_\theta)$ for each $\theta \in \Theta$. Then so is the form (2.1.6) on $L^2(R^n)$.
(ii) If each $\mathscr{E}^\theta$ is expressed as (2.1.2) on the hyperplane F_θ, then the form (2.1.6) is Markovian symmetric form possessing the local property.

(ii) is clear. To prove the first assertion, we prepare a simple but important lemma.

Lemma 2.1.1. Put $x' = (x_1, x_2, \ldots, x_{n-1})$ for $x = (x_1, \ldots, x_{n-1}, x_n)$, then

$$(2.1.7)\quad \int_{R^{n-1}} u(x', 0)^2 dx' \leq C\{\mathbf{D}(u, u) + (u, u)\},\quad u \in C_0^\infty(R^n),$$

C being a positive constant independent of u.

Proof. Take a C^∞-function $\phi(x_n)$ which is equal to 1 on a neighbourhood of $x_n = 0$ and vanishing when $x_n \geq 1$. Applying the Schwarz inequality to $u(x', 0) = -\int_0^1 \frac{\partial}{\partial x_n}\{\phi(x_n)u(x', x_n)\}\,dx_n$, we get

$$\int_{R^{n-1}} u(x', 0)^2 dx' \leq \int_{R^{n-1}} \left[\int_0^1 \left(\frac{\partial}{\partial x_n}\{\phi(x_n)u(x', x_n)\}\right)^2 dx_n\right] dx',$$

which yields (2.1.7). Q.E.D.

Proof of Theorem 2.1.3 (i). Consider $u_k \in C_0^\infty(R^n)$ such that $\mathscr{E}(u_k - u_l, u_k - u_l) \to 0$, $k, l \to \infty$, and $u_k \to 0$ in $L^2(R^n)$. Since the Dirichlet integral is closable, we have

$$\mathbf{D}(u_k, u_k) \to 0, \quad k \to \infty. \tag{2.1.8}$$

The above lemma and the invariance of the Dirichlet integral under the orthogonal transform imply

$$\int_{F\theta} u^\theta(x')^2 dx' \leqq C\{\mathbf{D}(u, u) + (u, u)\}, \quad u \in C_0^\infty(D), \tag{2.1.9}$$

where dx' is the $(n-1)$-dimensional Lebesgue measure on F_θ. Hence we get for each $\theta \in \Theta$

$$\lim_{k\to\infty} u_k^\theta = 0 \text{ on } L^2(F_\theta). \tag{2.1.10}$$

Choose a subsequence k_j such that $\sum_{j=1}^\infty \sqrt{\mathscr{E}(u_{k_j} - u_{k_{j+1}}, u_{k_j} - u_{k_{j+1}})} < \infty$. This means that $\sum_{j=1}^\infty \sqrt{\mathscr{E}^\theta(u_{k_j}^\theta - u_{k_{j+1}}^\theta, u_{k_j}^\theta - u_{k_{j+1}}^\theta)} < \infty$ for μ-a.e. $\theta \in \Theta$. In view of (2.1.10), we now have $\lim_{k_j\to\infty} \mathscr{E}^\theta(u_{k_j}^\theta, u_{k_j}^\theta) = 0$ for such θ because of the closability of $\mathscr{E}^\theta$. By making use of Fatou's lemma, $\mathscr{E}(u_m, u_m) \leqq \int_\theta \lim_{k_j\to\infty} \mathscr{E}^\theta(u_m^\theta - u_{k_j}^\theta, u_m^\theta - u_{k_j}^\theta)\mu(d\theta) + c\mathbf{D}(u_m, u_m) \leqq \lim_{k_j\to\infty} \mathscr{E}(u_m - u_{k_j}, u_m - u_{k_j}) + c\mathbf{D}(u_m, u_m)$, which can be made arbitrarily small by choosing m sufficiently large. Q.E.D.

For instance, the symmetric form on $L^2(R^2)$ defined by

$$\mathscr{E}(u, v) = c\cdot\mathbf{D}(u. v) + \int_{R^2} \frac{\partial u(x_1, x_2)}{\partial x_1}\frac{\partial v(x_1, x_2)}{\partial x_1} dx_1\mu(dx_2) + \int_{R^2} \frac{\partial u(x_1, x_2)}{\partial x_2}\frac{\partial v(x_1, x_2)}{\partial x_2}\nu(dx_1)\,dx_2 \tag{2.1.11}$$

is a special case of (2.1.2) with

$$\begin{cases} \nu_{11}(dx_1dx_2) = cdx_1dx_2 + dx_1\mu(dx_2) \\ \nu_{22}(dx_1dx_2) = cdx_1dx_2 + \nu(dx_1)dx_2 \\ \nu_{12}(dx_1dx_2) = \nu_{21}(dx_1dx_2) = 0. \end{cases}$$

Here μ and ν are any positive Radon measures on R^1. (2.1.11) is also of the type

(2.1.6) and we see on account of Theorem 2.1.3 that the form (2.1.11) is closable on $L^2(R^2)$.

We exhibit one more example of (2.1.2):

$$(2.1.12)\quad \mathscr{E}(u, v) = c\cdot\mathbf{D}(u, v) + \int_{-\infty}^{\infty}\int_{-\infty}^{\infty} \frac{\partial u(0, x_2, x_3)}{\partial x_2}\frac{\partial v(0, x_2, x_3)}{\partial x_2} dx_2 dx_3$$
$$+ \int_{-\infty}^{\infty}\int_{-\infty}^{\infty} \frac{\partial u(0, x_2, x_3)}{\partial x_3}\frac{\partial v(0, x_2, x_3)}{\partial x_3} dx_2 dx_3$$
$$+ \int_{-\infty}^{\infty} \frac{\partial u(0, 0, x_3)}{\partial x_3}\frac{\partial v(0, 0, x_3)}{\partial x_3} dx_3.$$

Applying Theorem 2.1.3 twice, we see that (2.1.12) is a closable form on $L^2(R^3)$.

(3°) One dimensional case.

Let ν be a positive Radon measure on R^1. Our present problem is as follows: Under what conditions on ν is the following form closable on $L^2(R^1)$?

$$(2.1.13)\quad \begin{cases} \mathscr{E}(u, v) = \int_{-\infty}^{\infty} u'(x)v'(x)\nu(dx) \\ \mathscr{D}[\mathscr{E}] = C_0^\infty(R^1). \end{cases}$$

Given a non-negative Borel function $a(x)$ on R^1, we say that $x \in R^1$ is a *regular point* of a if $\int_{(x-\varepsilon, x+\varepsilon)} \frac{1}{a(\xi)} d\xi$ is finite for some $\varepsilon > 0$. The set of all regular points of a is denoted by $R(a)$. The complementary set $R^1 - R(a)$ is denoted by $S(a)$ and called the *singular set* of a.

Theorem 2.1.4. The form (2.1.13) is closable on $L^2(R^1)$ if and only if the following conditions are satisfied:
(i) ν is absolutely continuous.
(ii) the density function a ($\nu(dx) = a(x)dx$) is locally integrable and vanishes a.e. on its singular set $S(a)$.

Proof. Necessity of (i). If condition (i) is not satisfied, then there exists a compact set K such that $\nu(K) > 0$, $|K| = 0$, $|\ \ |$ being the Lebesgue measure. Without loss of generality we assume $K \subset (0, 1)$. Take a sequence $\psi_n^0 \in C_0^\infty(R^1)$ such that Supp $[\psi_n^0] \subset (0,1)$ and ψ_n^0 decreases to the indicator function I_K as $n \uparrow \infty$. We set $\psi_n(x) = \psi_n^0(x) - \psi_n^0(x-1)$ and $\phi_n(x) = \int_{-\infty}^x \psi_n(x)dx$.

Then ϕ_n is a C^∞ function with Supp$[\phi_n] \subset (0, 2)$ and $\phi_n(x)^2 \leq$

$2\int_0^2 \psi_n(x)^2 dx$, $0 \leq x \leq 2$. Since ψ_n converges boundedly to $\psi(x) = I_K(x) - I_K(x-1)$, $(\phi_n, \phi_n) = \int_0^2 \phi_n(x)^2 dx \leq 4\int_0^2 \psi_n(x)^2 dx \to 0$, $n \to \infty$, whereas $\mathscr{E}(\phi_n, \phi_n) = \int_0^2 \psi_n(x)^2 \nu(dx)$ converges to a non-zero limit $\int_0^2 \psi(x)^2 \nu(dx)$. Obviously $\{\phi_n\}$ constitutes an $\mathscr{E}$-Cauchy sequence. Hence $\mathscr{E}$ is not closable.

Sufficiency of (ii). Suppose that ν has a density function a satisfying the condition (ii). Let $\phi_n \in C_0^\infty(R^1)$ constitute an $\mathscr{E}$-Cauchy sequence and be L^2-convergent to zero. Then $\psi_n = \phi_n'$ is $L^2(a\cdot dx)$ convergent to a function $\psi \in L^2(a \cdot dx)$.

For any finite interval $[\alpha, \beta) \subset R(a)$, observe the inequality

$$\left(\int_\alpha^\beta \psi(x)dx\right)^2 \leq 2(\phi_n(\beta) - \phi_n(\alpha))^2 + 2\left(\int_\alpha^\beta (\psi(x) - \psi_n(x))dx\right)^2.$$

The first term on the right hand side converges to zero for a.e. α and a.e. β. The second term converges to zero because it is dominated by $2\int_\alpha^\beta \frac{dx}{a(x)} \cdot \int_\alpha^\beta (\psi(x) - \psi_n(x))^2 a(x)dx$. Hence ψ vanishes a.e. on $R(a)$ and $a(x)dx$-a.e. on R^1 by condition (ii). We get $\lim_{n\to\infty} \mathscr{E}(\phi_n, \phi_n) = \int_{R^1} \psi(x)^2 a(x)dx = 0$ —the closability of $\mathscr{E}$. Q.E.D.

We omit the complicated part of the proof—the necessity of (ii). We refer the interested readers to M. Hamza [42] in this connection.

§2.2 Beurling-Deny Formulae

Let X and m be as in §1.1. We are concerned with a general repesentation theorem on Dirichlet forms first presented in Beurling and Deny [12].

Theorem 2.2.1. Any regular Dirichlet form $\mathscr{E}$ on $L^2(X;m)$ can be expressed for $u, v \in \mathscr{D}[\mathscr{E}] \cap C_0(X)$ as follows:

(2.2.1)
$$\mathscr{E}(u, v) = \mathscr{E}^{(c)}(u, v) + \int_{X\times X-d} (u(x) - u(y))(v(x) - v(y))J(dx, dy) + \int_X u(x)v(x)k(dx).$$

Here $\mathscr{E}^{(c)}$ is a symmetric form with domain $\mathscr{D}[\mathscr{E}^{(c)}] = \mathscr{D}[\mathscr{E}] \cap C_0(X)$

and satisfies the following condition stronger than the local property:

(2.2.2) $\mathscr{E}^{(c)}(u, v) = 0$ for $u \in \mathscr{D}[\mathscr{E}^{(c)}]$ and $v \in \vartheta(u)$,

where

(2.2.3) $\vartheta(u) = \{v \in \mathscr{D}[\mathscr{E}^{(c)}];\ v$ is constant on a neighbourhood of $\operatorname{Supp}[u]\}$.

J is a symmetric positive Radon measure on the product space $X \times X$ off the diagonal d and k is a positive Radon measure on X.

Such $\mathscr{E}^{(c)}$, J, and k are uniquely determined by $\mathscr{E}$. Furthermore, every normal contraction operates on the form $\mathscr{E}^{(c)}$.

We call the J and k occuring in the above theorem the *jumping measure* and the *killing measure*[1] respectively associated with the regular Dirichlet form $\mathscr{E}$.

Proof. (I) Uniqueness. It follows from (2.2.1) that

(2.2.4)
$$\mathscr{E}(u, v) = -2\int_{X\times X} u(x)v(y)J(dx, dy)$$

for any $u, v \in \mathscr{D}[\mathscr{E}] \cap C_0(X)$ with disjoint supports. In view of the regularity of $\mathscr{E}$ and Lemma 1.4.2, the functions $\sum_{i=1}^{p} u_i(x)v_i(y)$ such that $u_i, v_i \in \mathscr{D}[\mathscr{E}] \cap C_0(X)$, $\operatorname{Supp}[u_i] \cap \operatorname{Supp}[v_i] = \phi$, $1 \leq i \leq p$, constitute a dense subalgebra of $C_0(X \times X - d)$. Hence a positive Radon measure J on $X \times X - d$ satisfying (2.2.4) is uniquely determined by $\mathscr{E}$.

Next observe the equality

(2.2.5)
$$\mathscr{E}(u, v) = \int_{X\times X-d} (u(x) - u(y))(v(x) - v(y))J(dx, dy) + \int_X u(x)k(dx),$$

where v is a function in $\mathscr{D}[\mathscr{E}] \cap C_0(X)$ identically equal to 1 on a neighbourhood of $\operatorname{Supp}[u]$. Since such a v exists for any $u \in \mathscr{D}[\mathscr{E}] \cap C_0(X)$ by virtue of Theorem 1.4.2, (2.2.5) uniquely determines a positive Radon measure k.

As the measures J and k are unique, the form $\mathscr{E}^{(c)}$ of Theorem 2.2.1 is unique as well.

(II) The measure J as a vague limit. We utilize the approximating

1) Cf. Theorem 4.5.2.

form $\mathscr{E}^{(\beta)}$ of $\mathscr{E}$ defined by (1.3.16) together with the positive Radon measure σ_β on $X \times X$ determined by Lemma 1.4.1 and (1.4.7). Then

$$(2.2.6)\quad \mathscr{E}^{(\beta)}(u, v) = -\beta \int_{X\times X} u(x)v(y)\sigma_\beta(dxdy) \to \mathscr{E}(u, v), \quad \beta \to \infty$$

if $u, v \in \mathscr{D}[\mathscr{E}] \cap C_0(X)$ and $\mathrm{Supp}[u] \cap \mathrm{Supp}[v] = \phi$. This implies that there exists a unique positive Radon measure J on $X \times X - d$ satisfying (2.2.4) and

$$(2.2.7)\quad \frac{\beta}{2}\sigma_\beta \to J \quad \text{vaguely on } X \times X - \mathrm{d} \text{ as } \beta \to \infty.$$

In fact, we can see by (2.2.6) that the family $\{\beta\sigma_\beta, \beta > 0\}$ of measures is uniformly bounded on each compact subset of $X \times X - d$ and hence a subsequence $\beta_n\sigma_{\beta_n}$ converges as $\beta_n \to \infty$ vaguely on $X \times X - \mathrm{d}$ to a positive Radon measure J. Since J satisfies (2.2.4) and we have already observed the uniqueness of such J, we get (2.2.7).

(III) Construction of k and $\mathscr{E}^{(c)}$. Fix a metric ρ on X compatible with the given topology. Choose a sequence of relatively compact open sets G_l increasing to X and a sequence of numbers $\delta_l \downarrow 0$ such that the set $\Gamma_l = \{(x,y) \in G_l \times G_l; \rho(x, y) \geqq \delta_l\}$ is a continuous set with respect to the measure J for every l.

In the same way as (1.4.8) we can rewrite the approximating form $\mathscr{E}^{(\beta)}$ as follows: for $u \in \mathscr{D}[\mathscr{E}] \cap C_0(X)$ such that $\mathrm{Supp}\,[u] \subset G_l$,

$$(2.2.8)\quad \mathscr{E}^{(\beta)}(u, u) = \frac{1}{2}\beta \int_{G_l\times G_l} (u(x) - u(y))^2\sigma_\beta(dxdy)$$
$$+ \beta \int_{G_l} u(x)^2(1 - \beta G_\beta I_{G_l}(x))m(dx).$$

This particularly implies that the family of measures $\beta(1 - \beta G_\beta I_{G_l}(x))\cdot m(dx)$, $\beta > 0$, is uniformly bounded on each compact subset of G_l. Hence a subsequence $\beta_n \uparrow \infty$ and positive Radon measures k_l on G_l exist and for each l

$$(2.2.9)\quad \beta_n\,(1 - \beta_n G_{\beta_n} I_{G_l}) \cdot m \to k_l \text{ vaguely on } G_l,\ \beta_n \to \infty.$$

In view of (2.2.8), (2.2.9) and (2.2.7), we have for every l such that $\mathrm{Supp}[u] \subset G_l$,

$$(2.2.10)\quad \mathscr{E}(u, u) = \lim_{\beta_n\to\infty} \frac{\beta_n}{2} \int_{G_l\times G_l\,\cdot\,\rho(x\ y)<\delta_l} (u(x) - u(y))^2\sigma_{\beta_n}(dxdy)$$

$$+\int_{\Gamma_l}(u(x)-u(y))^2J(dxdy)+\int_{G_l}u(x)^2k_l(dx).$$

Extend the measures k_l to X by setting $k_l(E)=k_l(E\cap G_l)$; then by (2.2.9) k_l is non-increasing on each compact set. Denote the vague limit by k, i.e.,

(2.2.11) $k_l \to k$ vaguely on X, $l\to\infty$.

Letting l tend to infinity in (2.2.10), we arrive at the desired expression (2.2.1) with

(2.2.12) $$\mathscr{E}^{(c)}(u,v)=\lim_{l\to\infty}\lim_{\beta_n\to\infty}\frac{\beta_n}{2}\int_{G_l\times G_l,\rho(x,y)<\delta_l}(u(x)-u(y)) \times(v(x)-v(y))\sigma_{\beta_n}(dxdy).$$

This expression tells us that $\mathscr{E}^{(c)}$ satisfies the condition (2.2.2) and that every normal contraction operates on $\mathscr{E}^{(c)}$. Q.E.D.

The form $\mathscr{E}^{(c)}$ in Theorem 2.2.1 is called the *diffusion part* of the Dirichlet form $\mathscr{E}$.[1] When the underlying space X is equipped with a differential structure, the diffusion part can be expressed more explicitly. Indeed, we now show the form (1.2.1) exhibited in §1.2 is the most general expression of a closable Markovian symmetric form with domain $C_0^\infty(D)$.

Theorem 2.2.2. Consider the case where X is a domain D of R^n. Then any closable Markovian symmetric form $\mathscr{E}$ on $L^2(D;m)$ with $\mathscr{D}[\mathscr{E}]=C_0^\infty(D)$ can be expressed uniquely as (1.2.1).

Proof. By virtue of Theorem 2.1.1, the smallest closed extension $\mathscr{E}$ of $\mathscr{E}$ is a regular Dirichlet form on $L^2(D,m)$. Applying Theorem 2.2.1 to $\mathscr{E}$, we see that $\mathscr{E}$ is expressed as (2.2.1) for $u, v\in C_0^\infty(D)$ with unique measures J and k. Hence it suffices to derive the expression

(2.2.13) $$\mathscr{E}^{(c)}(u,v)=\sum_{i,j=1}^{n}\int_D\frac{\partial u(x)}{\partial x_i}\frac{\partial v(x)}{\partial x_j}\nu_{ij}(dx)$$

for $u, v\in C_0^\infty(D)$ with some Radon measures ν_{ij} satisfying (1.2.3).[2]

1) Cf. Theorem 4.5.2.
2) Cf. Theorem 5.4.4.

Let us prove (2.2.13) by making use of the approximation (2.2.12) of $\mathscr{E}^{(c)}$. Specifically,

(2.2.14) $\mathscr{E}^{(c)}(u, v)$

$$= \lim_{l\to\infty} \lim_{k\to\infty} \frac{\beta_k}{2} \int_{G\times G,\, |x-y|<\delta_l} (u(x) - u(y))\, (v(x) - v(y))\sigma_{\beta_k}(dx, dy).$$

The integral in the right hand side is independent, for sufficient small δ_l, of the choice of a relatively compact open set G containing the supports of u and v.

Let us fix relatively compact open sets G and G' such that $\bar{G} \subset G' \subset \bar{G}' \subset D$. Choose functions $f_1, f_2, \ldots, f_n \in C_0^\infty(D)$ such that

(2.2.15) $f_i(x) = x_i, \quad x \in G, \quad \mathrm{Supp}\,[f_i] \subset G', \quad 1 \leqq i \leqq n.$

Define measures $\{\nu_{ij}^{kl}\}$ on G' by

(2.2.16) $$\nu_{ij}^{kl}(E) = \frac{\beta_k}{2} \int_{\substack{E\times G'\\ |x-y|<\delta_l}} (f_i(x) - f_i(y))(f_j(x) - f_j(y))\sigma_{\beta_k}(dx, dy).$$

Since $\lim_{l\to\infty} \lim_{k\to\infty} \nu_{ii}^{kl}(G') = \mathscr{E}^{(c)}(f_i, f_i) \leqq \mathscr{E}(f_i, f_i) < \infty$ and the total variation of ν_{ij}^{kl} is dominated by $\sqrt{\nu_{ii}^{kl}(G')}\,\sqrt{\nu_{jj}^{kl}(G')}$, there exist bounded measures ν_{ij} on G' such that the convergence

(2.2.17) $\lim_{l\to\infty} \lim_{k\to\infty} \nu_{ij}^{kl} = \nu_{ij}$ vaguely, $1 \leqq i, \; j \leqq n,$

takes place by selecting subsequences of k and l if necessary.

The measures $\{\nu_{ij}\}$ satisfy the condition (1.2.3) on G′ because $\nu_{ij}^{kl} = \nu_{ji}^{kl}$ and $\sum_{i,j=1}^{n} \nu_{ij}^{kl}(E)\xi_i\xi_j = \frac{\beta_k}{2} \int_{E\times G'\, |x-y|<\delta_l} \{\sum_{i=1}^{n} \xi_i(f_i(x) - f_i(y))\}^2 \sigma_{\beta_k}(dx, dy).$ $\geqq 0$ for each k, l in view of (2.2.16).

We now expand a function $u \in C_0^\infty(D)$ with $\mathrm{Supp}\,[u] \subset G$ as $u(x) - u(y) = \sum_{i=1}^{n} \frac{\partial u(x)}{\partial x_i}(f_i(x) - f_i(y)) + \frac{1}{2} \sum_{i,j=1}^{n} \frac{\partial^2 u}{\partial x_i \partial x_j}(x + \theta(y - x)) \quad (f_i(x) - f_i(y))\,(f_j(x) - f_j(y)), 0 < \theta < 1,\ x, y \in G$. Then by (2.2.14) we see that $\mathscr{E}^{(c)}(u, u)$ is equal to $\lim_{l\to\infty} \lim_{k\to\infty} \{\sum_{i,j=1}^{n} \frac{\beta_k}{2} \int_{\substack{G\times G\\ |x-y|<\delta_l}} \frac{\partial u(x)}{\partial x_i} \frac{\partial u(x)}{\partial x_j}(f_i(x) - f_i(y))$ $(f_j(x) - f_j(y))\sigma_{\beta_k}(dx, dy) + I(k, l)\}$. Clearly the domain of the above integral can be replaced by $\{(x, y) \in G \times G';\ |x - y| < \delta_l\}$ for sufficiently large l. The error term $I(k, l)$ consists of integrals on the same domain, but each integrand has the product $(f_i(x) - f_i(y))\,(f_j(x) - f_j(y))$

$(f_p(x) - f_p(y))$, $1 \leqq i, j, p \leqq n$, as its factor. Hence it follows from (2.2.17) that $\lim_{l\to\infty} \lim_{k\to\infty} |I(k, l)| = 0$ and

$$(2.2.18)\quad \mathscr{E}^{(c)}(u, v) = \sum_{i,j=1}^{n} \int_G \frac{\partial u(x)}{\partial x_i} \frac{\partial v(x)}{\partial x_j} \nu_{ij}(dx).$$

Since G is arbitrary, (2.2.13) is proven. Q.E.D.

§2.3 Maximum Markovian Extensions

Consider an abstract real Hilbert space H with inner product (,) and a symmetric linear operator S on H with the domain $\mathscr{D}(S)$ being dense in H. We assume that $-S$ is non-negative definite: $(-Su,u) \geqq 0, u \in \mathscr{D}(S)$.

We are concerned with family

(2.3.1) $\mathscr{A}(S) = \{A; A$ is a self-adjoint extension of S and $-A$ is non-negative definite$\}$.

The family $\mathscr{A}(S)$ is non-empty. In fact,

$$(2.3.2)\quad \mathscr{E}_{(S)}(u, v) = (-Su, v), \mathscr{D}[\mathscr{E}_{(S)}] = \mathscr{D}(S)$$

defines a closable symmetric form on H on account of Problem 1.1.2. Hence the smallest closed extension $\bar{\mathscr{E}}_{(S)}$ gives rise to a self-adjoint operator, say A_0, by virtue of Theorem 1.3.1.

Lemma 2.3.1. (i) $A_0 \in \mathscr{A}(S)$.
(ii) If $A \in \mathscr{A}(S)$, then $\mathscr{E}_A$ is a closed extension of $\mathscr{E}_{(S)}$: $\mathscr{D}[\mathscr{E}_A] \supset \mathscr{D}[\mathscr{E}_{(S)}]$ and $\mathscr{E}_A(u, v) = \mathscr{E}_{(S)}(u, v)$, $u, v \in \mathscr{D}[\mathscr{E}_{(S)}]$. Here $\mathscr{E}_A$ denotes the closed symmetric form corresponding to A by Theorem 1.3.1.

Proof. (i) For any $u \in \mathscr{D}(S)$, we put $(\alpha - S)u = f$, $\alpha > 0$. Then $\mathscr{E}_{(S)}(u,v) + \alpha(u,v) = (f,v)$, $v \in \mathscr{D}[\mathscr{E}_{(S)}]$, because this relation is evident for $v \in \mathscr{D}(S)$. From this and (1.3.7), we see that $u = G_\alpha^0 f \in \mathscr{D}(A_0)$ and $(\alpha - A_0)u = f$, G_α^0 being the resolvent of A_0. Hence $\mathscr{D}(S) \subset \mathscr{D}(A_0)$ and $A_0 = S$ on $\mathscr{D}(S)$.
(ii) If $S \subset A$, then $\mathscr{D}[\mathscr{E}_{(S)}] = \mathscr{D}(S) \subset \mathscr{D}(A) \subset \mathscr{D}[\mathscr{E}_A]$ and $\mathscr{D}_{(S)}(u, v) = (-Su, v) = (-Au, v) = \mathscr{E}_A(u, v)$, $\forall u, v \in \mathscr{D}[\mathscr{E}_{(S)}]$, according to Corollary to Theorem 1.3.1. This means that $\mathscr{E}_A$ is also the closed extension of $\mathscr{E}_{(S)}$. Q.E.D.

The self-adjoint operator A_0 is called the *Friedrichs extension* of S. Let us introduce a semi-order $\prec$ in $\mathscr{A}(S)$ by

$$(2.3.3)\quad A_1 < A_2 \Leftrightarrow \mathscr{D}[\mathscr{E}_{A_1}] \subset \mathscr{D}[\mathscr{E}_{A_2}],\ \mathscr{E}_{A_1}(u, u) \geqq \mathscr{E}_{A_2}(u, u),\ {}^{\forall}u \in \mathscr{D}[\mathscr{E}_{A_1}].$$

Lemma 2.3.1 tells us that the Friendrichs extension A_0 is the minimum element of $\mathscr{A}(S)$ with the above semi-order, i.e., $A_0 < A$, ${}^{\forall}A \in \mathscr{A}(S)$. M. G. Krein found in [57] that there exists the maximum element, say A_K, among the family $\mathscr{A}(S)$, i.e., $A < A_K$, ${}^{\forall}A \in \mathscr{A}(S)$. We call A_K the *Krein extension* of S. We need a simple lemma before we can get an idea how to describe A_K explicitly.

Consider the resolvent $\{G_\alpha^0, \alpha > 0\}$ of the Friedrichs extension A_0. Since $(G_\alpha^0 u, u)$ is non-decreasing as $\alpha \downarrow 0$, the following symmetric form J^0 on H in the wide sense is well-defined:

$$(2.3.4)\quad J^0(u, v) = \lim_{\alpha \downarrow 0} (G_\alpha^0 u, v),\ \mathscr{D}[J^0] = \{u \in H;\ \lim_{\alpha \downarrow 0} (G_\alpha^0 u, u) < \infty\}.$$

Let S^* be the adjoint operator of S. We set

$$(2.3.5)\quad \mathscr{N}_\alpha = \{u \in \mathscr{D}(S^*);\ (\alpha - S^*)u = 0\},\ \mathscr{N}_\alpha^0 = \mathscr{N}_\alpha \cap \mathscr{D}[J^0]$$

for each $\alpha > 0$. We may interpret the space $\mathscr{N}_\alpha^0$ as the space of α-harmonic elements with finite (0-order) energy integrals.

Lemma 2.3.2. Let $-A$ be a non-negative definite self-adjoint operator on H.
(i) $A \in \mathscr{A}(S)$ if and only if $S^* \supset A$.
(ii) If $A \in \mathscr{A}(S)$, the Hilbert space $(\mathscr{D}[\mathscr{E}_A], \mathscr{E}_{A,\alpha})$ can be orthogonally decomposed as

$$(2.3.6)\quad \mathscr{D}[\mathscr{E}_A] = \mathscr{D}[\mathscr{E}_{(S)}] \oplus \mathscr{H}_{A,\alpha},\quad \alpha > 0,$$

with $\mathscr{H}_{A,\alpha} = \mathscr{N}_\alpha \cap \mathscr{D}[\mathscr{E}_A] = \mathscr{N}_\alpha^0 \cap \mathscr{D}[\mathscr{E}_A]$. Moreover, the following inequality holds:

$$(2.3.7)\quad \mathscr{E}_A(u, u) \geqq \alpha^2 J^0(u, u),\quad u \in \mathscr{H}_{A,\alpha}.$$

Proof. (i) If $S^* \supset A$, then $A \supset S^{**}$, while S^{**} is the smallest closed extension of S.[1)]
(ii) By virtue of Lemma 2.3.1, $\mathscr{D}[\mathscr{E}_{(S)}]$ is a closed subspace of $(\mathscr{D}[\mathscr{E}_A]$,

---

1) Cf. F.Riesz and B.Sz.Nagy [76; §117].

$\mathscr{E}_{A,\alpha}$). Denote by $\mathscr{H}_{A,\alpha}$ its orthogonal complement, i.e., $u \in \mathscr{H}_{A,\alpha}$ if and only if $u \in \mathscr{D}[\mathscr{E}_A]$ and $\mathscr{E}_{A,\alpha}(u, v) = 0$, $\forall v \in \mathscr{D}(S)$. Since $S \subset A$, $\mathscr{E}_{A,\alpha}(u, v)$ can then be written as $(u, (-S + \alpha)v)$ in view of Corollary to Theorem 1.4.1. Hence $\mathscr{H}_{A,\alpha} = \mathscr{N}^\alpha \cap \mathscr{D}[\mathscr{E}_A]$.

It only remains to prove the inequality (2.3.7). Note that the following transform $P_{\beta,\alpha}$ defines a one-to-one correspondence between the spaces $\mathscr{N}_\alpha$ and $\mathscr{N}_\beta$ $(\alpha,\beta > 0)$ and that $P_{\alpha,\beta} = P_{\beta,\alpha}^{-1}$:

$$(2.3.8) \quad P_{\beta,\alpha}u = u + (\alpha - \beta)G_\beta^0 u, \quad u \in \mathscr{A}_\alpha.$$

In fact, we see from (i) that $A_0 \subset S^*$ and $(\beta - S^*)G_\beta^0 u = u$. Hence $P_{\beta,\alpha}u \in \mathscr{D}(S^*)$ for $u \in \mathscr{N}_\alpha$ and $(\beta - S^*)P_{\beta,\alpha}u = (\beta - \alpha)u + (\alpha - \beta)u = 0$, that is, $P_{\beta,\alpha}u \in \mathscr{N}_\beta$. The resolvent equation immediately leads us to $P_{\alpha,\beta}P_{\beta,\alpha}u = u$, $u \in \mathscr{N}_\alpha$.

Now fix $\alpha > 0$ and $u \in \mathscr{H}_{A,\alpha}$ and put $u_\beta = P_{\beta,\alpha}u$. Using the orthogonality (2.3.6), we have $\mathscr{E}_{A,\alpha}(u, u) = \mathscr{E}_{A,\alpha}(u, u_\beta) = \mathscr{E}_{A,\beta}(u_\beta, u_\beta) + (\alpha - \beta)(u, u_\beta) \geqq (\alpha - \beta)(u, u_\beta) = (\alpha - \beta)(u, u) + (\alpha - \beta)^2(u, G_\beta^0 u)$. Therefore, $\mathscr{E}_A(u, u) \geqq (\alpha - \beta)^2(u, G_\beta^0 u) - \beta(u, u)$. Letting β tend to 0, we arrive at (2.3.7). Q.E.D.

Lemma 2.3.2 (ii) suggests that the following expression $\mathscr{E}_K$ is a candidate for the closed symmetric form associated with the Krein extension A_K:

$$(2.3.9) \quad \mathscr{D}[\mathscr{E}_K] = \mathscr{D}[\mathscr{E}_{(S)}] + \mathscr{N}_\alpha^0$$

$$(2.3.10) \quad \mathscr{E}_K(u, v) = \begin{cases} \mathscr{E}_{(S)}(u, v), & u, v \in \mathscr{D}[\mathscr{E}_{(S)}] \\ \alpha^2 J^0(u, v), & u, v \in \mathscr{N}_\alpha^0 \\ -\alpha(u, v), & u \in \mathscr{D}[\mathscr{E}_{(S)}], v \in \mathscr{N}_\alpha^0. \end{cases}$$

Actually, it can be shown that (2.3.9) and (2.3.10) define a closed symmetric form $\mathscr{E}_K$ on H independent of the choice of $\alpha > 0$. It is then easy to see by making use of Lemma 2.3.2 (i) that the self-adjoint operator A_K associated with $\mathscr{E}_K$ is an extension of S, namely, $A_K \in \mathscr{A}(S)$. Clearly this A_K must be the Krein extension of S in view of Lemma 2.3.2 (ii).

When S satisfies an additional condition

$$(2.3.11) \quad (-Su, u) \geqq \gamma(u, u), \quad \forall u \in \mathscr{D}(S),$$

for some constant $\gamma > 0$, the expressions (2.3.9) and (2.3.10) of $\mathscr{E}_K$ can be simplified as

(2.3.12) $\mathscr{D}[\mathscr{E}_K] = \mathscr{D}[\mathscr{E}_{(S)}] + \mathscr{N}_0$

(2.3.13) $\mathscr{E}_K(u, v) = \begin{cases} \mathscr{E}_{(S)}(u, v), u, & v \in \mathscr{D}[\mathscr{E}_{(S)}] \\ 0, & \text{otherwise.} \end{cases}$

Indeed, under the condition (2.3.11), $\lim_{\alpha \downarrow 0} (G_\alpha^0 u, u)$ is finite for any $u \in H$ and hence we arrive at (2.3.12) and (2.3.13) by letting α (formally) tend to 0 in (2.3.9) and (2.3.10) respectively. Moreover, in this case a direct description of the Krein extension A_K is possible:

(2.3.14) $\mathscr{D}(A_K) = \mathscr{D}(S^{**}) + \mathscr{N}_0$

(2.3.15) $A_K u = \begin{cases} S^{**}u, & u \in \mathscr{D}(S^{**}) \\ 0, & u \in \mathscr{N}_0. \end{cases}$

We shall not give a detailed proof of the above mentioned facts on $\mathscr{E}_K$ and A_K because our main concern is a certain subfamily of $\mathscr{A}(S)$ defined below rather than the whole family $\mathscr{A}(S)$.

Now let X and m be as in §1.1 and $-S$ be a non-negative definite symmetric linear operator on $L^2(X;m)$ such that the associated symmetric form $\mathscr{E}_{(S)}$ is Markovian. Define a subfamily $\mathscr{A}_M(S)$ of $\mathscr{A}(S)$ by

(2.3.16) $\mathscr{A}_M(S) = \{A \in \mathscr{A}(S)$: the semi-group on $L^2(X;m)$ generated by A is Markovian$\}$.

By virtue of Theorem 1.4.1, $A \in \mathscr{A}(S)$ belongs to $\mathscr{A}_M(S)$ if and only if $\mathscr{E}_A$ is a Dirichlet form on $L^2(X;m)$.

Theorem 2.1.1 implies that the Friedrichs extension A_0 of S is an element of $\mathscr{A}_M(S)$; indeed, it is the minimum one of $\mathscr{A}_M(S)$ with semi-order $\prec$ defined by (2.3.3). Natural questions then arise:

(I) Is there the maximum element, say A_R, in $\mathscr{A}_M(S)$?

(II) Does $A_R = A_k$?

We give answers to these questions in the simple case that X is a domain $D \subset R^n$, m is the Lebesgue measure and

(2.3.17) $S = \frac{1}{2}\Delta\left(= \frac{1}{2}\sum_{i=1}^{n} \frac{\partial^2}{\partial x_i^2}\right), \quad \mathscr{D}(S) = C_0^\infty(D).$

We then have

(2.3.18) $\mathscr{E}_{(S)}(u, v) = \frac{1}{2}\mathbf{D}(u, v), \quad \mathscr{D}[\mathscr{E}_{(S)}] = C_0^\infty(D).$

Recall the Sobolev spaces $H^1(D)$ and $H_0^1(D)$ in Example 1.2.3. As we saw in §1.2, $\mathscr{E}_{(S)}$ is a Markovian symmetric form and the smallest closed extension $\bar{\mathscr{E}}_{(S)}$ equals to $(1/2\ \mathbf{D}, H_0^1(D))$. The Friedrichs extension A_0 of S thus corresponds to the Dirichlet form $(1/2\,\mathbf{D}, H_0^1(D))$, while the larger Dirichlet form $(1/2\ \mathbf{D}, H^1(D))$ plays a striking role as we demonstrate below.

Theorem 2.3.1. Let A_R be the self-adjoint operator on $L^2(D)$ corresponding to the Dirichlet form $(1/2\ \mathbf{D}, H^1(D))$ (see Theorem 1.3.1). Then A_R is the maximum element of $\mathscr{A}_M(S)$.

The proof of this theorem is completed by the following two lemmas.

Lemma 2.3.3. $A_R \in \mathscr{A}_M(S)$.

Proof. Let $\{G_\alpha^R, \alpha > 0\}$ be the resolvent of A_R. Then, in view of (1.3.7), $u = G_\alpha^R f$, $f \in L^2(D)$, satisfies the equation

$$(2.3.19)\quad \left(\alpha - \frac{1}{2}\Delta\right)u = f$$

in the sense of Schwartz distribution. This equation is in turn equivalent to the statement $u \in \mathscr{D}(S^*)$, $(\alpha - S^*)u = f$ for S of (2.3.17). Therefore $A_R \subset S^*$ and hence $S \subset A_R$ by Lemma 2.3.2 (i), proving $A_R \in \mathscr{A}_M(S)$. Q.E.D.

Lemma 2.3.4. If $A \in \mathscr{A}_M(S)$, then
$\mathscr{D}[\mathscr{E}_A] \subset H^1(D)$ and $\mathscr{E}_A(u, u) \geqq 1/2\,\mathbf{D}(u, u)$, $\quad u \in \mathscr{D}[\mathscr{E}_A]$.

Proof. It suffices to show that for a fixed $\alpha > 0$,

$$(2.3.20)\quad \mathscr{E}_A(u,\ u) \geqq \frac{1}{2}\mathbf{D}(u,\ u),\quad {}^\forall u \in \mathscr{N}_\alpha \cap \mathscr{D}[\mathscr{E}_A].$$

In fact, by Lemma 2.3.2 (ii) any $u \in \mathscr{D}[\mathscr{E}_A]$ can be decomposed as $u = u_0 + u_h$, $u_0 \in H_0^1(D)$, $u_h \in \mathscr{N}_\alpha \cap \mathscr{D}[\mathscr{E}_A]$ and $\mathscr{E}_{A,\alpha}(u,\ u) = \mathbf{D}_\alpha(u_0, u_0) + \mathscr{E}_{A,\alpha}(u_h, u_h)$.[1] On the other hand, we have from (2.3.19)

$$(2.3.21)\quad \mathscr{N}_\alpha = \left\{u \in L^2(D);\ \left(\alpha - \frac{1}{2}\Delta\right)u = 0\right\}.$$

Hence, if (2.3.20) is true, then $\mathscr{E}_{A,\alpha}(u,\ u) \geqq \mathbf{D}_\alpha(u_0,\ u_0) + \mathbf{D}_\alpha(u_h,\ u_h) = \mathbf{D}_\alpha(u, u)$ by (1.2.17).

1) $\mathbf{D}_\alpha(u, u) = 1/2\,\mathbf{D}(u, u) + \alpha(u, u)$.

In order to prove (2.3.20), consider the resolvent $\{G_\alpha, \alpha > 0\}$ on $L^2(D)$ of A. Since $A \subset S^*$ by lemma 2.3.2 (i), the function $u = G_\alpha f, f \in C_0^\infty(D)$, satisfies the distribution equation (2.3.19). αG_α being assumed to be Markovian, we can use Weyl's lemma[1] to get a Markovian resolvent kernel $\{R_\alpha(x, E), \alpha > 0\}$ on D such that

$$(2.3.22) \quad G_\alpha f = R_\alpha f \text{ a.e. on } D, \quad f \in C_0^\infty(D).$$

Then notice that the distribution equation (2.3.19) is valid for f and $u = R_\alpha f$ whenever $|f|$ and $R_\alpha |f|$ are locally integrable. Since in this case $(R_\beta |f|, |\phi|) = (|f|, R_\beta |\phi|) \to 0, \beta \to \infty, {}^\forall \phi \in C_0^\infty(D)$, we see $\lim_{\beta\to\infty} (\beta R_\beta f, \phi) = (f, \phi) + \lim_{\beta\to\infty} (R_\beta f, 1/2\Delta\phi) = (f, \phi), {}^\forall \phi \in C_0^\infty(D)$. This implies that the relation

$$(2.3.23) \quad \lim_{\beta\to\infty} \beta(f - \beta R_\beta f) = -\frac{1}{2}\Delta f$$

is valid in the sense of Schwartz distribution when $|f|$ and $R_\alpha|f|$ are locally integrable.

We now proceed to the proof of (2.3.20) with the help of the relation (2.3.23). Take any $u \in \mathcal{N}_\alpha \cap \mathcal{D}[\mathcal{E}_A]$. By Weyl's lemma again, we may assume that u is infinitely differentiable and satisfies the equation $(\alpha - 1/2\Delta)\, u(x) = 0$, $x \in D$, in the ordinary sense. Writing the approximating form $\mathcal{E}_A^{(\beta)}(u, u)$ of $\mathcal{E}_A(u, u)$ in the manner of (1.4.8), we get

$$(2.3.24) \quad \begin{cases} \mathcal{E}_A^{(\beta)}(u, u) = \dfrac{1}{2}(f_\beta, 1) \\ f_\beta(x) = \beta^2 R_\beta(u(x) - u)^2(x) + 2\beta u(x)^2(1 - \beta R_\beta 1(x)). \end{cases}$$

Since f_β can be rewritten as $f_\beta = -\beta(u^2 - \beta R_\beta u^2) + 2\beta u(u - \beta R_\beta u) + \beta u^2(1 - \beta R_\beta 1)$ and u is infinitely differentiable, the relation (2.3.23) leads us to

$$(2.3.25) \quad \lim_{\beta\to\infty} \frac{1}{2}(f_\beta, \phi) = (\frac{1}{4}\Delta u^2 - \alpha u^2, \phi), \quad \phi \in C_0^\infty(D).$$

In particular, $1/4\,\Delta u^2 - \alpha u^2 \geqq 0$ because $f_\beta \geqq 0, {}^\forall \beta > 0$, by (2.3.24).

On account of (2.3.24) and (2.3.25), we have, for any $\phi \in C_0^\infty(D)$ such that $0 \leqq \phi \leqq 1$, $\mathcal{E}_A(u, u) = \lim_{\beta\to\infty} \mathcal{E}_A^{(\beta)}(u, u) \geqq (1/4)\,\Delta u^2 - \alpha u^2, \phi)$. Letting

1) Cf. footnote 2) of p. 11.

then $\phi \uparrow 1$, $\mathscr{E}_A(u, u) \geqq (1/4)\ \Delta u^2 - \alpha u^2, 1) = (1/4)(\Delta u^2, 1) - \alpha(u, u) = 1/2\ \mathbf{D}(u, u)$, completing the proof of (2.3.20). Q.E.D.

Suppose the domain $D \subset R^n$ is bounded; then Poincaré's inequality (1.5.8) holds and hence S of (2.3.17) satisfies the condition (2.3.11). Therefore, the closed symmetric form $\mathscr{E}_K$ corresponding to the Krein extension A_K of S possesses the property that $\mathscr{N}_0 \subset \mathscr{D}[\mathscr{E}_K]$, $\mathscr{E}_K(u, u) = 0$, $\forall u \in \mathscr{N}_0$ in view of (2.3.12) and (2.3.13). Here $\mathscr{N}_0 = \{u \in L^2(D) : \Delta u = 0\}$. Since there is a harmonic function with nonzero finite Dirichlet integral, we see $\mathscr{E}_K \neq (1/2\mathbf{D}, H^1(D))$ and hence $A_K \neq A_R$. In particular, A_K does not generate a Markovian semi-group because otherwise $A_K \prec A_R$ and $A_K = A_R$.

Theorem 2.3.2. If the domain $D \subset R^n$ is bounded, the semigroup on $L^2(D)$ generated by the Krein extension A_K of the operator (2.3.17) is not Markovian.

We can look upon the family $\mathscr{A}(S)$ from a different view-point. $A \in \mathscr{A}(S)$ is, in general, a restriction of S^* by Lemma 2.3.2 (i). Hence A may be expressed in the following manner:

$$(2.3.26) \quad \begin{cases} \mathscr{D}(A) = \{u \in \mathscr{D}(S^*) : u \text{ satisfies } L\} \\ Au = S^*u, \quad u \in \mathscr{D}(A). \end{cases}$$

Here L stands for the *lateral condition* to specify the restriction. In many cases, L can be described in terms of the behaviour of u near the boundary and L is then called the *boundary condition*.

Let us examine the simplest case that $D = (a, b), -\infty < a < b < \infty$, $Su = 1/2u''$, $\mathscr{D}(S) = C_0^\infty((a, b))$. As we easily see,

$$(2.3.27) \quad \begin{cases} \mathscr{D}(S^*) = \{u \in L^2((a, b)) : u \text{ is differentiable, } u' \text{ is absolutely} \\ \qquad\qquad \text{continuous and } u'' \in L^2((a, b))\} \\ S^*u = 1/2u''. \end{cases}$$

The following fact is known:[1]

$$(2.3.28) \quad \mathscr{D}(S^{**}) = \{u \in \mathscr{D}(S^*) : u(a) = u'(a) = u(b) = u'(b) = 0\}.$$

Since $\mathscr{D}(A_K) = \mathscr{D}(S^{**}) + \mathscr{N}_0$ by (2.3.14) and $\mathscr{N}_0$ consists of linear functions, we get

1) Cf. T.Kato [54; Chap.III, Example 5.32].

(2.3.29) $\mathscr{D}(A_K) = \mathscr{D}(S^*) \cap \{u\colon u \text{ satisfies } L_K\}$

with

$$L_K \begin{cases} -u'(a) - \dfrac{1}{b-a}u(a) + \dfrac{1}{b-a}u(b) = 0 \\ u'(b) + \dfrac{1}{b-a}u(a) - \dfrac{1}{b-a}u(b) = 0. \end{cases}$$

Namely, the Krein extension A_K is decided by the boundary condition L_K. As we have already seen in Example 1.3.2, the Friedrichs extension A_0 and the maximum element A_R of $\mathscr{A}_M(S)$ can be described by the following boundary conditions L_0 and L_R respectively:

$$L_0\colon \quad u(a) = u(b) = 0$$

$$L_R\colon \quad u'(a) = u'(b) = 0.$$

It is known in the present case that each element A of $\mathscr{A}(S)$ can be characterized by a boundary condition. One of the general types of the boundary condition is the following:[1]

$$L: \quad \begin{cases} -u'(a) + \pi_{11}u(a) + \pi_{12}u(b) = 0 \\ u'(b) + \pi_{21}u(a) + \pi_{22}u(b) = 0, \quad \pi_{12} = \pi_{21}. \end{cases}$$

Let $\{T_t, t > 0\}$ be the semigroup generated by A. Then the following criteria hold:[2]

T_t is positive $\Leftrightarrow \pi_{12} \leqq 0$

T_t is Markovian $\Leftrightarrow \pi_{12} \leqq 0,\ \pi_{i1} + \pi_{i2} \leqq 0,\ i = 1, 2.$

L_R corresponds to $\pi_{ij} = 0$, while L_K corresponds to $-\pi_{11} = -\pi_{22} = \pi_{12} = \dfrac{1}{b-a} > 0$. Hence the semigroup generated by A_K is not even positive.

1) W.Feller [33].
2) W.Feller [33].

CHAPTER 3

Potential Theory of Dirichlet Forms

This chapter is devoted to the presentation of the potential theory of regular Dirichlet forms due to Beurling and Deny. Its connection with classical potential theory is stated in the examples.

Fundamental notions in the potential theory (sets of capacity zero, equilibrium potentials, quasi-continuity of functions, etc.) shall be interpreted probabilistically in §4.3. New notions of regular nests (§3.1) and smooth measures (§3.2) shall play important roles in Chapter 6 and Chapter 5, respectively. A theorem of Beurling-Deny on spectral synthesis (§3.3) shall be utilized in §4.4 and §5.3.

§3.1 Capacity and Quasi-Continuity

Throughout Chapter 3 we fix a pair (X, m) of §1.1 and consider a regular Dirichlet form $\mathscr{E}$ on $L^2(X; m)$. As in §1.5, we denote $\mathscr{D}[\mathscr{E}]$ by $\mathscr{F}$. Although we present the potential theory relevant to the Hilbert space $(\mathscr{F}, \mathscr{E}_1)$, an analogous presentation is possible in terms of the extended Dirichlet space $(\mathscr{F}_e, \mathscr{E})$ provided that $(\mathscr{F}, \mathscr{E})$ is transient. There is no essential difference in these two presentations, however, as we shall see at the end of this section.

Denote by $\mathscr{O}$ the family of all open subsets of X. For $A \in \mathscr{O}$ we define

$$(3.1.1) \quad \mathscr{L}_A = \{u \in \mathscr{F};\ u \geqq 1 \ m\text{-a.e. on } A\}$$

$$(3.1.2) \quad \mathrm{Cap}\,(\mathrm{A}) = \begin{cases} \inf\limits_{u \in \mathscr{L}_A} \mathscr{E}_1(u, u), & \mathscr{L}_A \neq \phi \\ \infty & \mathscr{L}_A = \phi, \end{cases}$$

and for any set $A \subset X$ we let

$$(3.1.3) \quad \mathrm{Cap}\,(A) = \inf_{B \in \mathscr{O}, A \subset B} \mathrm{Cap}\,(B).$$

We call this *1-capacity* of A or simply the *capacity* of A.

Let $\mathcal{O}_0 = \{A \in \mathcal{O};\ \mathcal{L}_A \neq \phi\}$. The function e_A in the following lemma is called a *1-equilibrium potential* of the set $A \in \mathcal{O}_0$.

Lemma 3.1.1. (i) For each $A \in \mathcal{O}_0$, there exists a unique element $e_A \in \mathcal{L}_A$ such that

$$(3.1.4) \quad \mathcal{E}_1(e_A, e_A) = \mathrm{Cap}\,(A).$$

(ii) $0 \leqq e_A \leqq 1$ m-a.e. and $e_A = 1$ m-a.e. on A.
(iii) e_A is a unique element of $\mathcal{F}$ satisfying $e_A = 1$ m-a.e. on A and $\mathcal{E}_1(e_A, v) \geqq 0$, $\forall v \in \mathcal{F}$, $v \geqq 0$ m-a.e. on A.
(iv) $v \in \mathcal{F}$, $v = 1$ m-a.e. on $A \Rightarrow \mathcal{E}_1(e_A, v) = \mathrm{Cap}\,(A)$.
(v) $A, B \in \mathcal{O}_0$, $A \subset B \Rightarrow e_A \leqq e_B$ m-a.e.

Proof. (i) Clearly $\mathcal{L}_A$ is a convex closed subset of $(\mathcal{F}, \mathcal{E}_1)$. By making use of the equality

$$(3.1.5) \quad \mathcal{E}_1\left(\frac{u-v}{2}, \frac{u-v}{2}\right) + \mathcal{E}_1\left(\frac{u+v}{2}, \frac{u+v}{2}\right) = \frac{1}{2}\mathcal{E}_1(u, u) + \frac{1}{2}\mathcal{E}_1(v, v)$$

we can see that any minimizing sequence $u_n \in \mathcal{L}_A$ ($\lim_{n\to\infty} \mathcal{E}_1(u_n, u_n) = \mathrm{Cap}(A)$) is $\mathcal{E}_1$-convergent to an element $e_A \in \mathcal{L}_A$ satisfying (3.1.4) and that such an e_A is unique.
(ii) By $(\mathcal{E}.4)'$, $u = (0 \vee e_A) \wedge 1 \in \mathcal{L}_A$ and $\mathcal{E}_1(u, u) \leqq \mathcal{E}_1(e_A, e_A) = \mathrm{Cap}\,(A)$. Hence $u = e_A$.
(iii) If v has the stated property, then $e_A + \varepsilon v \in \mathcal{L}_A$ and $\mathcal{E}_1(e_A + \varepsilon v, e_A + \varepsilon v) \geqq \mathcal{E}_1(e_A, e_A)$ for any $\varepsilon > 0$. From this we get $\mathcal{E}_1(e_A, v) \geqq 0$. Conversely, suppose $u \in \mathcal{F}$ satisfies the conditions in (iii), then $u \in \mathcal{L}_A$ and $w - u \geqq 0$ m-a.e. on A for any $w \in \mathcal{L}_A$. Hence $\mathcal{E}_1(w, w) = \mathcal{E}_1(u + (w-u), u + (w-u)) \geqq \mathcal{E}_1(u, u)$, $\forall w \in \mathcal{L}_A$, proving $u = e_A$.
(iv) Immediate from (iii).
(v) Notice that $\mathcal{E}_1(u^+, u^-) \leqq 0$ for any $u \in \mathcal{F}$. Since $e_A - e_A \wedge e_B = (e_A - e_B)^+$ vainishes m-a.e. on A, $\mathcal{E}_1(e_A - e_A \wedge e_B, e_A - e_A \wedge e_B) = \mathcal{E}_1(-e_A \wedge e_B, (e_A - e_B)^+) = \mathcal{E}_1((e_A - e_B)^-, (e_A - e_B)^+) - \mathcal{E}_1(e_B, (e_A - e_B)^+) \leqq 0$ and hence $e_A = e_A \wedge e_B$ m-a.e. Q.E.D.

The capacity defined by (3.1.2) has the following properties on the family $\mathcal{O}$.

Lemma 3.1.2. (i) $A, B \in \mathcal{O}, A \subset B \Rightarrow \mathrm{Cap}\,(A) \leqq \mathrm{Cap}\,(B)$.
(ii) $\mathrm{Cap}\,(A \cup B) + \mathrm{Cap}\,(A \cap B) \leqq \mathrm{Cap}\,(A) + \mathrm{Cap}\,(B),\ A, B \in \mathcal{O}$.
(iii) $A_n \in \mathcal{O}, A_n \uparrow \Rightarrow \mathrm{Cap}\,(\bigcup_n A_n) = \sup_n \mathrm{Cap}\,(A_n)$.

Proof. (i) is clear. For $A, B \in \mathcal{O}_0$, $\mathrm{Cap}\,(A \cup B) + \mathrm{Cap}\,(A \cap B) \leqq \mathscr{E}_1(e_A \vee e_B, e_A \vee e_B) + \mathscr{E}_1(e_A \wedge e_B, e_A \wedge e_B) = 1/2\, \mathscr{E}_1(e_A + e_B, e_A + e_B) + 1/2\, \mathscr{E}_1(|e_A - e_B|,\ |e_A - e_B|) \leqq \mathscr{E}_1(e_A,\ e_A) + \mathscr{E}_1(e_B,\ e_B) = \mathrm{Cap}\,(A) + \mathrm{Cap}\,(B)$, which proves (ii). It suffices to prove (iii) when $\sup_n \mathrm{Cap}\,(A_n)$ is finite. By Lemma 3.1.1, we then have for $n > m$ $\mathscr{E}_1(e_{A_n} - e_{A_m}, e_{A_n} - e_{A_m}) = \mathrm{Cap}\,(A_n) - \mathrm{Cap}\,(A_m)$. Therefore, e_{A_n} is $\mathscr{E}_1$-convergent to some $u \in \mathscr{F}$. Clearly $u = 1$ m-a.e. on $A = \bigcup_n A_n$. If $v \in \mathscr{F}$ is non-negative m-a.e. on A, $\mathscr{E}_1(u, v) = \lim_{n\to\infty} \mathscr{E}_1(e_{A_n}, v) \geqq 0$. Hence $u = e_A$ by Lemma 3.1.1 and $\sup_n \mathrm{Cap}\,(A_n) = \lim_{n\to\infty} \mathscr{E}_1(e_{A_n}, e_{A_n}) = \mathscr{E}_1(u,\ u) = \mathrm{Cap}(A)$. Q.E.D.

As a consequence of Lemma 3.1.2, we can obtain the next theorem.[1]

Theorem 3.1.1. The capacity defined by (3.1.2) and (3.1.3) is a *Choquet capacity*, i.e.,
(i) $A \subset B \Rightarrow \mathrm{Cap}\,(A) \leqq \mathrm{Cap}\,(B)$
(ii) $A_n \uparrow \Rightarrow \mathrm{Cap}\,(\bigcup_n A_n) = \sup_n \mathrm{Cap}\,(A_n)$
(iii) A_n compact, $A_n \downarrow \Rightarrow \mathrm{Cap}\,(\bigcap_n A_n) = \inf_n \mathrm{Cap}\,(A_n)$.

According to a well-known theorem on the capacitability of the Choquet capacity,[2] it holds that for any Borel set A

$$(3.1.6)\quad \mathrm{Cap}\,(A) = \sup_{K:\text{compact},\ K \subset A} \mathrm{Cap}\,(K).$$

The present notion of capacity enables us to think of exceptional sets finer than m-negligible sets. In fact, the inequality $m(A) \leqq \mathrm{Cap}\,(A)$ for $A \in \mathcal{O}$ implies that any set of zero capacity is m-negligible. Let A be a subset of X. A statement depending on $x \in A$ is said to hold *q.e.* on A if there exists a set $N \subset A$ of zero capacity such that the statement is true for every $x \in A - N$. "q.e." is an abbreviation of *"quasi-everywhere"*.

Let $X_\Delta = X \cup \Delta$ be the one-point compactification of X. When X is already compact, Δ is regarded as an isolated point. For any subset $A \subset$

1) Cf. P.A.Meyer [65].
2) Cf. P.A.Meyer [65].

X, $A \cup \Delta$ is endowed with the relative topology as a subspace of X_Δ. Any function u on A is always extended to a function on $A \cup \Delta$ by setting $u(\Delta) = 0$. In particular, any function belonging to $C_\infty(X)$ is regarded as a continuous function on X_Δ.

Let u be a function defined q.e. on X. We call u *quasi-continuous* if there exists for any $\varepsilon > 0$ an open set $G \subset X$ such that $\mathrm{Cap}\,(G) < \varepsilon$ and $u|_{X-G}$ is continuous. Here $u|_{X-G}$ denotes the restriction of u to $X - G$. If we require a stronger condition by replacing $u|_{X-G}$ in the above definition with $u|_{X_\Delta - G}$, then we say that u is *quasi-continuous in the restricted sense*.

A sequence $\{F_k\}$ of closed sets such that $F_k \uparrow$ and $\mathrm{Cap}\,(X - F_k) \downarrow 0$, $k \uparrow \infty$, is called a *nest* on X. A nest $\{F_k\}$ is said to be *m-regular* if for each k, $\mathrm{Supp}\,(I_{F_k} \cdot m) = F_k$, i.e., $m(U(x) \cap F_k) \neq 0$ for any $x \in F_k$ and any open neighbourhood $U(x)$ of x.

Lemma 3.1.3. For a given nest $\{F_k\}$ on X, let $F'_k = \mathrm{Supp}\,(I_{F_k} \cdot m)$. Then $\{F'_k\}$ is an m-regular nest.

Proof. F'_k is closed and $F'_k \subset F_k$. Moreover,

$$(3.1.7) \quad m(F_k - F'_k) = 0.$$

In fact, by setting $G_k = X - F_k$, $G'_k = X - F'_k$, we see $G_k \subset G'_k = \{x \in X: m(U(x) - G_k) = 0, \ {}^\exists U(x)\}$. By the Lindelöf covering theorem, there exist points $x_l \in G'_k$ such that $G'_k = \bigcup_l U(x_l)$. Hence $m(G'_k - G_k) \leqq \sum_l m(U(x_l) - G_k) = 0$.

(3.1.7) implies $\mathscr{L}_{G'_k} = \mathscr{L}_{G_k}$ and, consequently, $\mathrm{Cap}\,(G'_k) = \mathrm{Cap}\,(G_k) \to 0$, $k \to \infty$, proving that $\{F'_k\}$ is again a nest. $\{F'_k\}$ is m-regular because, for any $x \in F'_k$ and for any $U(x)$, $m(U(x) \cap F'_k) \geqq m(U(x) \cap F_k) - m(F_k - F'_k) > 0$ by virtue of (3.1.7). Q.E.D.

Given a nest $\{F_k\}$ on X, let

$$(3.1.8) \quad C(\{F_k\}) = \{u\colon u|_{F_k} \text{ is continuous for each } k\}$$

$$(3.1.9) \quad C_\infty(\{F_k\}) = \{u\colon u|_{F_k \cup \Delta} \text{ is continuous for each } k\}.$$

Clearly $C_\infty(\{F_k\}) \subset C(\{F_k\})$, $C_\infty(X) \subset C_\infty(\{F_k\})$ and $C(X) \subset C(\{F_k\})$.

Theorem 3.1.2. (i) Let S be a countable family of quasi-continuous functions (resp. quasi-continuous functions in the restricted sense) on X.

Then there exists a regular nest $\{F_k\}$ on X such that $S \subset C(\{F_k\})$ (resp. $S \subset C_\infty(\{F_k\})$).
(ii) Let $\{F_k\}$ be a regular nest on X and u belong to $C(\{F_k\})$. If $u \geqq 0$ m-a.e., then $u(x) \geqq 0$, $\forall x \in \bigcup_{k=1}^{\infty} F_k$.

Proof. (i) If u is quasi-continuous, there exists a sequence $\{\tilde{F}_k\}$ of closed sets such that $\text{Cap}\,(X - \tilde{F}_k) < 1/k$ and $u|_{\tilde{F}_k}$ is continuous. Let $F_k = \bigcup_{l=1}^{k} \tilde{F}_l$, $k = 1, 2, \ldots$, then $\{F_k\}$ is a nest and $u \in C(\{F_k\})$.

Given a family $S = \{u_l\}$ of quasi-continuous functions u_l, choose for each l a nest $\{F_k^{(l)}\}$ such that $u_l \in C(\{F_k^{(l)}\})$, $\text{Cap}\,(X - F_k^{(l)}) < (1/2^l)(1/k)$. Let $F_k = \bigcap_{l=1}^{\infty} F_k^{(l)}$. Since $\text{Cap}\,(X - F_k) \leqq \sum_{l=1}^{\infty} \text{Cap}\,(X - F_k^{(l)}) \leqq 1/k$, $\{F_k\}$ is a nest again. Clearly $S \subset C(\{F_k\})$. It suffices now to regularize $\{F_k\}$ according to Lemma 3.1.3.
(ii) Suppose $u(x) < 0$ for some $x \in F_k$, then there is a neighbourhood $U(x)$ such that $u(y) < 0$, $\forall y \in U(x) \cap F_k$, on account of the continuity of $u|_{F_k}$. Since $\{F_k\}$ is m-regular, $m(U(x) \cap F_k) > 0$, contradicting the assumption that $u \geqq 0$ m-a.e. Q.E.D.

In particular, the preceding theorem implies the following: if u is quasi-continuous and $u \geqq 0$ m-a.e., then $u \geqq 0$ q.e. This assertion can be localized to an arbitrary open set $G \subset X$. In fact, thinking of G instead of X, the notion of quasi-continuity of a function on G can be defined; furthermore, we can make assertions analogous to Lemma 3.1.3 and Theorem 3.1.2. In particular, we have

Lemma 3.1.4. Let G be an open set of X and u be quasi-continuous on G. If $u \geqq 0$ m-a.e. on G, then $u \geqq 0$ q.e. on G.

So far in this section we have not used the regularity of the Dirichlet form $\mathscr{E}$. We now state two theorems based on this property of $\mathscr{E}$. Given two functions u and v, v is said to be a *quasi-continuous modification* of u (*in the restricted sense*) if v is *quasi-continuous* (*in the restricted sense*) and $v = u$ m-a.e.

Theorem 3.1.3. Each $u \in \mathscr{D}[\mathscr{E}]$ admits a quasi-continuous modification in the restricted sense.

Proof. First we establish the inequality

(3.1.10) $$\text{Cap}\,\{x \in X: |u(x)| > \lambda\} \leqq \frac{1}{\lambda^2}\mathscr{E}_1(u, u),\ \lambda > 0,\ u \in \mathscr{F} \cap C(X).$$

Since $G = \{x \in X: |u(x)| > \lambda\} \in \mathcal{O}$ and $|u|/\lambda \in \mathcal{L}_G$ for any $\lambda > 0$ and $u \in \mathcal{F} \cap C(X)$, we have $\mathrm{Cap}(G) \leq (1/\lambda^2)\ \mathcal{E}_1(|u|, |u|) \leq (1/\lambda^2)\,\mathcal{E}_1(u, u)$, proving (3.1.10).

In view of the regularity of $\mathcal{E}$, any $u \in \mathcal{F}$ can be approximated with respect to the $\mathcal{E}_1$-metric by some $u_n \in \mathcal{F} \cap C_0(X)$. We may assume $\mathcal{E}_1(u_{k+1} - u_k, u_{k+1} - u_k) < 2^{-3k}$ by selecting a subsequence if necessary. Then $\mathrm{Cap}\,(G_k) \leq 2^{-k}$ by (3.1.10) where $G_k = \{x \in X: |u_{k+1}(x) - u_k(x)| > 2^{-k}\}$. Let $F_k = \bigcap_{l=k}^{\infty} G_l^c$. Clearly $\{F_k\}$ is a nest and $|u_n(x) - u_m(x)| \leq \sum_{\nu=N+1}^{\infty} |u_{\nu+1}(x) - u_\nu(x)| \leq 2^{-N}$ for any $x \in F_k$ and any $n, m > N \geq k$. This means, for each k, $u_n|_{F_k \cup \Delta}$ (by setting $u_n(\Delta) = 0$) are convergent uniformly as $n \to \infty$. Let $\tilde{u}(x) = \lim_{n\to\infty} u_n(x)$, $x \in \bigcup_{k=1}^{\infty} F_k$; then $\tilde{u} \in C_\infty(\{F_k\})$ and $u = \tilde{u}$ m-a.e. Q.E.D.

Denote by $\tilde{\mathcal{F}}$ the set of all quasi-continuous functions in the restricted sense belonging to $\mathcal{F}$. By the above theorem and Lemma 3.1.4, the equivalence classes of $\tilde{\mathcal{F}}$ (in the sense of q.e.) are identical with the equivalence classes of $\mathcal{F}$ (in the sense of m-a.e.).

Lemma 3.1.5. Inequality (3.1.10) holds for any $u \in \tilde{\mathcal{F}}$.

Proof. For $u \in \tilde{\mathcal{F}}$, consider $u_n \in \mathcal{F} \cap C_0(X)$ and its limit $\tilde{u}$ which appeared in the proof of Theorem 3.1.3. Since $u = \tilde{u}$ q.e., there is for any $\varepsilon > 0$ an open set G such that $\mathrm{Cap}\,(G) < \varepsilon$ and u_n converges to u uniformly on $X - G$. Hence, for any λ, ε_1 with $\lambda > \varepsilon_1 > 0$, $\{x \in X: |u(x)| > \lambda\} \subset \{x \in X: |u_n(x)| > \lambda - \varepsilon_1\} \cup G$ if n is sufficiently large. Then, by (3.1.10), $\mathrm{Cap}\,\{x \in X: |u(x)| > \lambda\} \leq \dfrac{\mathcal{E}_1(u_n, u_n)}{(\lambda - \varepsilon_1)^2} + \varepsilon$. By letting $n \to \infty$, $\varepsilon_1 \to 0$ and then $\varepsilon \to 0$, we get the inequality (3.1.10) for $u \in \tilde{\mathcal{F}}$. Q.E.D.

Theorem 3.1.4. (i) If $u_n \in \tilde{\mathcal{F}}$ constitutes an $\mathcal{E}_1$-Cauchy sequence, then there are a subsequence n_k and a $u \in \tilde{\mathcal{F}}$ such that $u_{n_k} \to u$ q.e. and $\{u_n\}$ is $\mathcal{E}_1$-convergent to u.
(ii) If $u_n \in \mathcal{F}$ constitutes an $\mathcal{E}_1$-Cauchy sequence and if some modifications $\tilde{u}_n \in \tilde{\mathcal{F}}$ of u_n converge to a function $\tilde{u}$ q.e., then $\tilde{u} \in \tilde{\mathcal{F}}$ and $\{u_n\}$ is $\mathcal{E}_1$-convergent o $\tilde{u}$.

Proof. (ii) follows from (i) at once. Suppose that $u_n \in \tilde{\mathcal{F}}$ and $\mathcal{E}_1(u_n - u_m, u_n - u_m) \to 0$. By making use of Lemma 3.1.5, we can see exactly as in the proof of Theorem 3.1.3 that there exist a subsequence n_l

and a decreasing sequence of sets $\{E_k\}$ such that $\lim_{k\to\infty} \mathrm{Cap}\,(E_k) = 0$ and u_{n_l} converges uniformly on each $X_\Delta - E_k$ as $n_l \to \infty$. Let u be the limit function. Choose for any $\varepsilon > 0$ an open set G_1 such that $\mathrm{Cap}\,(G_1) < \varepsilon/2$ and $G_1 \supset E_k$ for sufficiently large k. Also choose an open set G_2 such that $\mathrm{Cap}\,(G_2) < \varepsilon/2$ and $u_{n_l}|_{X_\Delta - G_2}$ is continuous for every n_l, which is possible in view of Theorem 3.1.2 (i).

Let $G = G_1 \cup G_2$. Then $\mathrm{Cap}(G) < \varepsilon$ and u_{n_l} converges to u uniformly on $X_\Delta - G$. Hence $u \in \mathscr{F}$. Clearly $\{u_n\}$ is also $\mathscr{E}_1$-convergent to u. Q.E.D.

We could have employed the metric $\mathscr{E}_\alpha$ for positive α instead of $\mathscr{E}_1$ in the definition (3.1.2) of capacity. However, it would make no difference as far as the notions of "q.e." and "quasi-continuity" are concerned because of the inequality $(1 \wedge \alpha)\,\mathscr{E}_1(u, u) \leqq \mathscr{E}_\alpha\,(u, u) \leqq (1 \vee \alpha)\,\mathscr{E}_1\,(u, u)$, $u \in \mathscr{F}$.

Suppose now that the Dirichlet space $(\mathscr{F}, \mathscr{E})$ is transient in the sense of §1.5. $(\mathscr{F}, \mathscr{E})$ is then completed to the extended Dirichlet space $(\mathscr{F}_e, \mathscr{E})$ by Theorem 1.5.2. $\mathscr{F}_e$ is a real Hilbert space with inner product $\mathscr{E}$ and

(3.1.11) $\displaystyle\int_X |u(x)|\,g(x)m(dx) \leqq \sqrt{\mathscr{E}(u, u)}, \ \forall u \in \mathscr{F}_e,$

g being a reference function of $(\mathscr{F}, \mathscr{E})$ (see §1.5).

Let us define the 0-order capacity $\mathrm{Cap}_{(0)}(A)$ by replacing $\mathscr{F}$ and $\mathscr{E}_1$ in (3.1.1) and (3.1.2) with $\mathscr{F}_e$ and $\mathscr{E}$ respectively. Clearly

(3.1.12) $\mathrm{Cap}_{(0)}\ (A) \leqq \mathrm{Cap}\ (A), \quad A \subset X,$

but $\mathrm{Cap}_{(0)}\ (A) = 0$ still implies $m(A) = 0$ on account of (3.1.11). Moreover, by making use of the above mentioned properties of $(\mathscr{F}_e, \mathscr{E})$ we can readily restate all the statements established in this section in terms of $(\mathscr{F}_e, \mathscr{E})$ and $\mathrm{Cap}_{(0)}$. In particular, we have the (0-order) equilibrium potential $e_A^{(0)}$ corresponding to the modified version of Lemma 3.1.1.

Lemma 3.1.6. Let $\{A_n\}$ be a decreasing sequence of relatively compact open sets. Then $\mathrm{Cap}_{(0)}\ (A_n) \downarrow 0$ if and only if $\mathrm{Cap}\ (A_n) \downarrow 0$.

Proof. The "if" part is trivial from (3.1.12). Suppose that $\mathrm{Cap}_{(0)}\ (A_n) \downarrow 0$. Then by the modified version of (3.1.4) and the inequality (3.1.11),

(3.1.13) $e_{A_n}^{(0)} \to 0 \quad m\text{-a.e.}$

by selecting a subsequence if necessary.

Since $\bar{A}_1$ is compact and $\mathscr{E}$ is regular, there is a non-negative function $g \in \mathscr{F} \cap C_0(X)$ such that $g(x) \geqq 1$, $x \in A_1$. We let $e'_n = e^{(0)}_{A_n} \wedge g$. Then $e'_n \in L^2(X;m) \cap \mathscr{F}_e = \mathscr{F}$ by Theorem 1.5.2 (ii). It is easy to see that $\mathscr{E}_1(e'_n, e'_n)$ is uniformly bounded by $\mathrm{Cap}_{(0)}(A_1) + \mathscr{E}_1(g, g)$. Hence the Cesàro means f_n of some subsequence of $\{e'_n\}$ are $\mathscr{E}_1$-convergent.

On the other hand, (3.1.13) implies that e'_n converges to zero m-a.e. and so does f_n. Therefore, $\lim_{n\to\infty} \mathscr{E}_1(f_n, f_n) = 0$. We now conclude $\mathrm{Cap}(A_n) \to 0$ because $f_n \in \mathscr{L}_{A_n}$ and $\mathrm{Cap}(A_n) \leqq \mathscr{E}_1(f_n, f_n)$. Q.E.D.

Theorem 3.1.5. (i) $\mathrm{Cap}_{(0)}(A) = 0$ if and only if $\mathrm{Cap}(A) = 0$ for $A \subset X$.
(ii) A function is quasi-continuous with respect to $\mathrm{Cap}_{(0)}$ if and only if it is quasi-continuous with respect to Cap.

Proof. (i) is clear from Lemma 3.1.6. The "if" part of (ii) is also clear from (3.1.12). Let $\{G_n\}$ be relatively compact open sets which converge to X. If u is quasi-continuous with respect to $\mathrm{Cap}_{(0)}$, then it is also quasi-continuous on each G_n with respect to Cap by virtue of Lemma 3.1.6. Then it is easy to see that u is quasi-continuous on X. Q.E.D.

Example 3.1.1 (BLD functions).[1] Suppose that the Dirichlet space (1/2**D**, $H^1_0(D)$) in Example 1.2.3 is transient in the sense of §1.5. This happens, for instance, when D is a bounded domain of R^n or when $D = R^n$ and $n \geqq 3$ as in Example 1.5.1 and Example 1.5.2.

Let $(\mathscr{F}_e, \mathscr{E})$ be the associated extended Dirichlet space. We may consider the space $\tilde{\mathscr{F}}_e$ of all quasi-continuous modifications of elements of $\mathscr{F}_e$ in accordance with the 0-order version of Theorem 3.1.3. Each element of $\tilde{\mathscr{F}}_e$ is called a *BLD (Beppo-Levi-Deny) function of potential type.* In view of Theorem 3.1.4, we can see that $u \in \tilde{\mathscr{F}}_e$ if and only if there is a sequence $u_n \in C_0^\infty(D)$ such that $\{u_n\}$ constitutes a Cauchy sequence with respect to the Dirichlet integral and u_n converges to u q.e. on D.

Consider the case that $D = R^n$ and $n \geqq 3$. The 0-order capacity $\mathrm{Cap}_{(0)}$ is then just the classical Newtonian capacity (see Example 3.3.1). Since $\mathscr{F}_e$ is contained in the space $\mathscr{G}$ of (1.5.18), each function of $\mathscr{F}_e$ has a version such as in the second paragraph of Example 1.2.3. But each element of $\tilde{\mathscr{F}}_e$ has a stronger property, namely, that it is absolutely continuous on almost all straight lines parallel to each axis no matter how the axes are taken.

§3.2 Measures of Finite Energy Integrals

A positive Radon measure μ on X is said to be *of finite energy integral* if

1) Cf. J. Deny-J. L. Lions [26] and J. L. Doob [28].

(3.2.1) $\int_X |v(x)|\mu(dx) \leq C\sqrt{\mathscr{E}_1(v,v)}, \quad v \in \mathscr{F} \cap C_0(X),$

for some positive constant C. A positive Radon measure μ on X is of finite energy integral if and only if there exists for each $\alpha > 0$ a unique function $U_\alpha\mu \in \mathscr{F}$ such that

(3.2.2) $\mathscr{E}_\alpha(U_\alpha\mu,\ v) = \int_X v(x)\mu(dx), \quad \forall v \in \mathscr{F} \cap C_0(X).$

We call $U_\alpha\mu$ an *α-potential.*

Let $\{T_t, t > 0\}$ be the Markovian semi-group on $L^2(X; m)$ associated with the Dirichlet form $\mathscr{E}$. $u \in L^2(X; m)$ is called *α-excessive* (with respect to $\{T_t, t > 0\}$) if

(3.2.3) $u \geq 0,\ e^{-\alpha t}T_t u \leq u, \quad m\text{-a.e.}, \ \forall t > 0.$

Theorem 3.2.1. The following conditions are equivalent to each other for $u \in \mathscr{F}$ and $\alpha > 0$.

(i) u is an α-potential.

(ii) u is α-excessive.

(iii) $u \geq 0$, $\beta G_{\beta+\alpha}u \leq u$, m-a.e., $\forall\beta > 0$, where G_β is the resolvent of T_t.

(iv) $\mathscr{E}_\alpha(u, v) \geq 0$, $\forall v \in \mathscr{F}$, $v \geq 0$ m-a.e.

(v) $\mathscr{E}_\alpha(u, v) \geq 0$, $\forall v \in \mathscr{F} \cap C_0(X)$, $\quad v \geq 0$.

Proof. The implication (ii) $\Rightarrow$ (iii) is clear from (1.3.1). By (1.3.18) and the resolvent equation, we have $\mathscr{E}_\alpha(u, v) = \lim_{\beta\to\infty} \beta(u - \beta G_{\beta+\alpha}u,\ v)$ which gives the implication (iii) $\Rightarrow$ (iv). Suppose that (iv) is satisfied. Then we can see by using (3.1.5) that u is the unique element in the convex set $\mathscr{L}_u = \{w \in \mathscr{F}: w \geq u\ m\text{-a.e.}\}$ minimizing $\mathscr{E}_\alpha(w, w)$. Since $|u| \in \mathscr{L}_u$, $u = |u| \geq 0$. Moreover, $(u - e^{-\alpha t}T_t u,\ v) = (u,\ v - e^{-\alpha t}T_t v) = \mathscr{E}_\alpha(u, G_\alpha v - e^{-\alpha t}T_t G_\alpha v)$, which is non-negative for any non-negative $v \in L^2(X; m)$ because $G_\alpha v - e^{-\alpha t}T_t G_\alpha v = \int_0^t e^{-\alpha s}T_s v ds$ is then non-negative as well. Hence u is α-excessive.

(i) $\Rightarrow$ (v) and (iv) $\Rightarrow$ (v) are trivial. Conversely, suppose that (v) is satisfied. Choose for any non-negative $v \in \mathscr{F}$ a sequence $v_n \in \mathscr{F} \cap C_0(X)$ $\mathscr{E}_1$-convergent to v. According to Theorem 1.4.2, v_n^+ is then $\mathscr{E}_1$-convergent to v as well and $\mathscr{E}_\alpha(u, v) = \lim_{n\to\infty} \mathscr{E}_\alpha(u, v_n^+) \geq 0$, which proves (iv).

Assume (v) and let $I(v) = \mathscr{E}_\alpha(u, v)$, $v \in \mathscr{F} \cap C_0(X)$. Consider any compact set K and choose non-negative $v_K \in \mathscr{F} \cap C_0(X)$ such that $v_K \geq 1$ on K. Then $|I(v)| \leq \|v\|_\infty \cdot I(v_K)$ for any $v \in \mathscr{F} \cap C_0(X)$ with $\mathrm{Supp}[v] \subset K$. Since any function $w \in C_0(X)$ can be approximated uniformly on X by functions $v \in \mathcal{F} \cap C_0(X)$ with $\mathrm{Supp}[v] \subset \mathrm{Supp}[w]$ by virtue of Lemma 1.4.2, the above inequality assures that I can be extended uniquely to a positive linear functional on $C_0(X)$. Hence u is an α-potential of a unique positive Radon measure, and so we obtain (i). Q.E.D.

Corollary. If $u_1, u_2 \in \mathcal{F}$ are α-potentials, then so are $u_1 \wedge u_2$ and $u_1 \wedge 1$.

This assertion is indeed trivial for α-excessive functions.

Problem 3.2.1. Let $\mathcal{D}$ be a core of $\mathcal{E}$ satisfying the following condition: $\mathcal{D}$ is a subalgebra of $\mathcal{F} \cap C_0(X)$ and there exists for any $\varepsilon > 0$ a function $\phi_\varepsilon(t)$ satisfying (1.1.5) such that $\phi_\varepsilon(u) \in \mathcal{D}$, $\forall u \in \mathcal{D}$. Show that each condition of Theorem 3.2.1 is also equivalent to the following weaker one:

$$(3.2.4) \quad \mathcal{E}_\alpha(u, v) \geqq 0, \ \forall v \in \mathcal{D}, v \geqq 0.$$

(Hint: derive (v) from this by modifying the proof of Theorem 2.1.2).

Problem 3.2.2. Let $\mathcal{D}$ be the same as in the above problem. Show that a positive Radon measure μ on X is of finite energy integral if and only if inequality (3.2.1) holds for any $v \in \mathcal{D}$.

We denote by S_0 the family of all positive Radon measures of finite energy integrals. We show that each $\mu \in S_0$ can be approximated by certain measures $\mu_n \in S_0$ which are absolutely continuous with respect to m.

Lemma 3.2.1. For $\mu \in S_0$ and $\alpha > 0$, let

$$(3.2.5) \quad g_n = n(U_\alpha\mu - nG_{n+\alpha}(U_\alpha\mu)), \quad n = 1, 2, \cdots.$$

Then $g_n \cdot m$ converges vaguely to μ and $G_\alpha g_n$ converges $\mathcal{E}_\alpha$-weakly to $U_\alpha\mu$.

Proof. Applying Theorem 3.2.1 to $u = U_\alpha\mu$, we see that $g_n \geqq 0$ m-a.e. Moreover, $\mathcal{E}_\alpha(G_\alpha g_n, v) = (g_n, v) = n(u - nG_{n+\alpha}u, v) \to \mathcal{E}_\alpha(u, v)$, $n \to \infty$, $v \in \mathcal{F}$. In particular, $\lim_{n\to\infty} (g_n, v) = \int_X v(x)\mu(dx)$, $\forall v \in \mathcal{F} \cap C_0(X)$. Q.E.D.

Lemma 3.2.2. Each measure in S_0 charges no set of zero capacity.

Proof. It suffices to prove for $\mu \in S_0$

$$(3.2.6) \quad \mu(G) \leqq \sqrt{\mathcal{E}_1(U_1\mu, U_1\mu)} \cdot \sqrt{\mathrm{Cap}(G)}, \ \forall G \in \mathcal{O}_0.$$

This follows from the preceding Lemma: $\mu(G) \leqq \varliminf_{n\to\infty} \int_G g_n(x)m(dx) \leqq$

$\lim_{n\to\infty} (g_n, e_G) = \mathscr{E}_1(U_1\mu, e_G) \leq \sqrt{\mathscr{E}_1(U_1\mu, U_1\mu)}\sqrt{\mathscr{E}_1(e_G, e_G)}$ which equals the right hand side of (3.2.6) in view of (3.1.4). Q.E.D.

Theorem 3.2.2. For any $\mu \in S_0$

(3.2.7) $\mathscr{F} \subset L^1(X; \mu)$

(3.2.8) $\mathscr{E}_\alpha(U_\alpha\mu, v) = \int_X \tilde{v}(x)\mu(dx), \quad \alpha > 0, \quad v \in \mathscr{F}.$

Here $\tilde{v}$ denotes any quasi-continuous modification (in the restricted sense) of $v \in \mathscr{F}$.

Proof. For any $v \in \mathscr{F}$, choose a sequence $v_n \in \mathscr{F} \cap C_0(X)$ which is $\mathscr{E}_1$-convergent to v. By Theorem 3.1.4 and Lemma 3.1.4, a subsequence n_k exists such that v_{n_k} converges q.e. on X to $\tilde{v}$, $\tilde{v}$ being any fixed quasi-continuous modification of v. Since the inequality (3.2.1) holds for every v_n, we have from the preceding lemma and Fatou's lemma that $\int_X |\tilde{v}(x) - v_n(x)|\mu(dx) = \int_X \varliminf_{n_k\to\infty} |v_{n_k}(x) - v_n(x)|\mu(dx) \leq C\cdot \varliminf_{n_k\to\infty} \sqrt{\mathscr{E}_1(v_{n_k} - v_n, v_{n_k} - v_n)}$, which implies that $\tilde{v} \in L^1(X; \mu)$ and v_n is $L^1(X; \mu)$-convergent to $\tilde{v}$. By letting n tend to infinity in equation (3.2.2) with v being replaced by v_n, we can arrive at (3.2.8). Q.E.D.

For $\mu \in S_0$ we set

(3.2.9) $\mathscr{E}_\alpha(\mu) = \mathscr{E}_\alpha(U_\alpha\mu, U_\alpha\mu).$

We may now call this the *α-energy integral* of μ because we have $\mathscr{E}_\alpha(\mu) = \int_X \widetilde{U_\alpha\mu}(x)\mu(dx)$ from Theorem 3.2.2. Theorem 3.2.2 also enables us to formulate the following maximum principle.

Lemma 3.2.3. Consider the two α-potentials $u_1 = U_\alpha\mu$ and $u_2 = U_\alpha\nu$, $\alpha > 0$, $\mu,\nu \in S_0$.

(i) If $\tilde{u}_1 \leq \tilde{u}_2$ μ-a.e., then $u_1 \leq u_2$ m-a.e.

(ii) If $\tilde{u}_1 \leq C$ μ-a.e. for some constant C, then $u_1 \leq C$ m-a.e.

Proof. (i) Set $u = u_1 \wedge u_2$. Then by (3.2.8)

$$\mathscr{E}_\alpha(u_1, u) = \int_X \tilde{u}(x)\mu(dx) = \int_X \tilde{u}_1(x)\mu(dx) = \mathscr{E}_\alpha(u_1, u_1).$$

Since u is also an α-potential and $u \leqq u_1$ m-a.e., $\mathscr{E}_\alpha(u-u_1, u-u_1) = \mathscr{E}_\alpha(u, u-u_1) - \mathscr{E}_\alpha(u_1, u-u_1) \leqq 0$, proving $u_1 = u \leqq u_2$ m-a.e.
(ii) It suffices to set $u_2 = u_1 \wedge C$. Q.E.D.

Let us call a (positive) Borel measure μ on X *smooth* if it satisfies the following conditions:
(S.1) μ charges no set of zero capacity
(S.2) there exists an increasing sequence $\{F_n\}$ of compact sets such that

(3.2.10) $\mu(F_n) < \infty, \ n = 1, 2, \ldots$

(3.2.11) $\mu(X - \bigcup_{n=1}^{\infty} F_n) = 0$

(3.2.12) $\lim_{n\to\infty} \mathrm{Cap}\,(K - F_n) = 0$ for any compact set K.

We denote by S the family of all smooth measures. The class S is quite large and it contains all positive Radon measures on X charging no set of zero capacity. In particular, $S \supset S_0$ in view of Lemma 3.2.2. We claim, however, that any measure in S can be approximated by measures in S_0. First, we need two lemmas.

Lemma 3.2.4. Let ν be a bounded positive Borel measure on X. If $\nu(A) \leqq C \cdot \mathrm{Cap}\,(A)$ for any Borel set A and for a positive constant C, then $\nu \in S_0$.

Proof. For any non-negative $v \in \mathscr{F} \cap C_0(X)$ with $\mathscr{E}_1(v, v) = 1$, we have by using (3.1.10)

$$\int_X v(x)\nu(dx) \leqq \nu(X) + \sum_{k=0}^{\infty} 2^{k+1}\nu\,\{x: 2^k \leqq v(x) < 2^{k+1}\} \leqq \nu(X)$$

$$+ C\sum_{k=0}^{\infty} 2^{k+1}\,\mathrm{Cap}(\{x: v(x) \geqq 2^k\}) \leqq \nu(X) + C\sum_{k=0}^{\infty} 2^{k+1}2^{-2k} = \nu(X) + 4C,$$

which shows that $\nu \in S_0$. Q.E.D.

Lemma 3.2.5. Let ν be a bounded positive Borel measure on X charging no set of zero capacity. Then there exists a decreasing sequence $\{G_n\}$ of open sets such that

(3.2.13) $\mathrm{Cap}(G_n) \to 0, \quad \nu(G_n) \to 0, \quad n \to \infty$

(3.2.14) $\nu(A) \leqq 2^n\,\mathrm{Cap}(A)$ for any Borel set $A \subset X - G_n$.

Proof. Fix n and let $\alpha = \inf\{2^n \operatorname{Cap}(A) - \nu(A): A$ is a Borel subset of $X\}$. Clearly $-\nu(X) \leqq \alpha$. If $\alpha < 0$, choose an open set B_1 such that $2^n\operatorname{Cap}(B_1) - \nu(B_1) \leqq \alpha/2$. Then $\alpha_1 = \inf\{2^n\operatorname{Cap}(A) - \nu(A): A \subset X - B_1\}$ is not less than $\alpha/2$ because $\alpha \leqq 2^n\operatorname{Cap}(A \cup B_1) - \nu(A \cup B_1) \leqq \{2^n \operatorname{Cap}(A) - \nu(A)\} + \{2^n \operatorname{Cap}(B_1) - \nu(B_1)\}$. If $\alpha_1 < 0$, choose a relatively open set $B_2 \subset X - B_1$ such that $2^n\operatorname{Cap}(B_2) - \nu(B_2) \leqq \alpha_1/2$. Continuing in this fashion we find open sets $\{B_1 \cup B_2 \cup \ldots B_k\}$ such that

$$(3.2.15)\quad \begin{cases} 2^n \operatorname{Cap}(B_1 \cup \ldots \cup B_k) - \nu(B_1 \cup \ldots \cup B_k) \leqq 0 \\ 2^n \operatorname{Cap}(A) - \nu(A) \geqq 2^{-k}\alpha, \forall A \subset X - (B_1 \cup \ldots \cup B_k). \end{cases}$$

Let us put $G_n' = \bigcup_{k=1}^{\infty} B_k$. Then by (3.2.15) $2^n\operatorname{Cap}(A) \geqq \nu(A)$ for any Borel subset A of $X - G_n'$. Furthermore, $2^n\operatorname{Cap}(G_n') \leqq \nu(G_n') \leqq \nu(X) < \infty$. Then it suffices to set $G_n = \bigcup_{m=n}^{\infty} G_m'$, because $\{G_n\}$ is a decreasing sequence of open sets with $\operatorname{Cap}(G_n) \leqq 2^{-n+1}\nu(X) \to 0$, $n \to \infty$. Since ν charges no set of zero capacity, $0 = \nu(\bigcap_n G_n) = \lim_{n\to\infty} \nu(G_n)$. Q.E.D.

Theorem 3.2.3. A positive Borel measure μ on X is smooth if and only if there exists a sequence $\{F_n\}$ of increasing closed sets satisfying (3.2.11) and (3.2.12) and $I_{F_n}\cdot\mu \in S_0$ for each n.

Proof. The "if" part is clear; indeed, it suffices to replace F_n by $F_n \cap \bar{E}_n$, $\{E_n\}$ being relatively compact open sets such that $\bar{E}_n \subset E_{n+1}$, $E_n \uparrow X$. The "only if" part follows from the preceding two lemmas. In fact, when μ is a bounded positive Borel measure charging no set of zero capacity, it is sufficient to adopt as F_n the complementary set of G_n of Lemma 3.2.5.

For a general measure $\mu \in S$, let $\{E_l\}$ be the associated compact sets in the condition (S.2). Since $\mu_l = I_{E_l}\cdot\mu$ is bounded for each l, μ_l admits a sequence $\{F_n^{(l)}\}$ with the properties in the statement of Theorem 3.2.3. We set $F_n = \bigcup_{l=1}^{n} \{E_l \cap F_n^{(l)}\}$, $n = 1, 2, \cdots$. $\{F_n\}$ are then increasing closed sets such that $I_{F_n}\cdot\mu \in S_0$ for each n. (3.2.11) and (3.2.12) for $\{F_n\}$ follow from the inclusion $K - F_n \subset (K - E_l) \cup (K - F_n^{(l)})$ for any compact set K. Q.E.D.

Finally, suppose that $(\mathscr{F}, \mathscr{E})$ is transient and let $(\mathscr{F}_e, \mathscr{E})$ be the extended Dirichlet space. We say that a positive Radon measure μ on X is *of finite* (*0-order*) *energy integral* if the inequality (3.2.1) holds with $\mathscr{E}_1$ on the right hand side being replaced by $\mathscr{E}$. Then the equation (3.2.2) with $\alpha = 0$ determines a unique function $U\mu \in \mathscr{F}_e$. This is called *the* (0-*order*) *potential* of μ.

It is clear that all arguments made so far in this section then extend to the case that $\alpha = 0$. In particular, the (0-order) energy integral $\mathscr{E}(\mu) = \mathscr{E}(U\mu, U\mu)$ equals $\int_X \widetilde{U\mu}(x)\mu(dx)$, $\widetilde{U\mu}$ being a quasi-continuous version of $U\mu$ with respect to $\mathrm{Cap}_{(0)}$ (or, equivalently, with respect to Cap by virtue of Theorem 3.1.5).

Example 3.2.1 (Riesz potentials). When $\alpha < n$, the space (1.4.24) with $\psi(x) = c|x|^\alpha$ is the Dirichlet space on $L^2(R^n)$ of the transient symmetric stable convolution semi-group of index α. According to Example 1.5.2, the extended Dirichlet space $(\mathscr{F}_e, \mathscr{E})$ is expressed as (1.5.14).

We can then assert that a positive Radon measure μ on R^n is of finite (0-order) energy integral if and only if

$$(3.2.16) \quad \int_{R^n}\int_{R^n} R^{(\alpha)}(x-y)\mu(dy)\mu(dx) < \infty,$$

where $R^{(\alpha)}$ is the Riesz kernel (1.5.15). Also, the *Riesz potential* $R^{(\alpha)}\mu(x) = \int_{R^n} R^{(\alpha)}(x-y)\,\mu(dy)$ is a quasi-continuous and lower semi-continuous version of the (0-order) potential $U\mu$ and the left hand side of (3.2.16) coincides with the (0-order) energy integral $\mathscr{E}(\mu)$ in this case.

If $\mu(dx) = f(x)dx$ with a non-negative bounded measurable f of compact support, then these assertions follow from Theorem 1.5.3, (1.5.10), (1.5.16), and the continuity of the potential $R^{(\alpha)}f(x) = \int_{R^n} R^{(\alpha)}(x-y)f(y)dy$. Suppose that μ satisfies the finiteness condition (3.2.16). In a manner similar to the proof of Lemma 3.2.1, we can construct a sequence of functions f_n of the above type such that $R^{(\alpha)}f_n(x)$ increases to $R^{(\alpha)}\mu(x)$ for each $x \in R^n$, and, furthermore, $R^{(\alpha)}f_n \in \mathscr{F}_e$ is $\mathscr{E}$-convergent to a function $u \in \mathscr{F}_e$. Consequently, $R^{(\alpha)}\mu$ is a quasi-continuous and lower semi-continuous version of u by virtue of Theorem 3.1.4. On account of the 0-order version of Theorem 3.2.1, u is a 0-order potential of a measure which must be identical with μ because $\mathscr{E}(u, R^{(\alpha)}g) = \lim_{n\to\infty} \mathscr{E}(R^{(\alpha)}f_n, R^{(\alpha)}g) = \lim_{n\to\infty}(R^{(\alpha)}f_n, g) = (R^{(\alpha)}\mu, g) = \int_{R^n} R^{(\alpha)}g(x)\mu(dx)$ for any g of the above type.

§ 3.3 Equilibrium Potentials and Reduced Functions

For $v \in \mathscr{F}$, we denote by $\tilde{v}$ a quasi-continuous modification of v in the restricted sense. We note that a property of $v \in \mathscr{F}$ such as $\tilde{v} \geq 0$ q.e. on A does not depend on the choice of the quasi-continuous version v.

Lemma 3.3.1. The following conditions are equivalent for $u \in \mathscr{F}$ and a closed set $F \subset X$:

(i) $u = U_\alpha \mu$, $\mu \in S_0$, with $\mathrm{Supp}[\mu] \subset F$.

(ii) $\mathscr{E}_\alpha(u, v) \geqq 0\ \forall v \in \mathscr{F}$, $\tilde{v} \geqq 0$ q.e. on F.

(iii) $\mathscr{E}_\alpha(u, v) \geqq 0\ \forall v \in \mathscr{F} \cap C_0(X)$, $v \geqq 0$ on F.

Proof. (i) $\Rightarrow$ (ii) is an immediate consequence of Theorem 3.2.2. (ii) $\Rightarrow$ (iii) is trivial. The implication (iii) $\Rightarrow$ (i) is clear as well from Theorem 3.2.1 and Lemma 1.4.2. Q.E.D.

Consider now a Borel set B with $\mathrm{Cap}\,(B) < \infty$.

$$(3.3.1)\quad \mathscr{L}_B = \{u \in \mathscr{F} : \tilde{u} \geqq 1 \quad \text{q.e.} \quad \text{on } B\}$$

is then a non-empty convex closed subset of $(\mathscr{F}, \mathscr{E}_1)$. Hence by the same reason as in the proof of Lemma 3.1.1, $\mathscr{L}_B$ admits a unique element $e_B \in \mathscr{L}_B$ minimizing $\mathscr{E}_1(u, u)$ on $\mathscr{L}_B$. Moreover, e_B is a unique element of $\mathscr{F}$ satisfying

$$(3.3.2)\quad \tilde{e}_B = 1 \quad \text{q.e.} \quad \text{on } B$$

$$(3.3.3)\quad \mathscr{E}_1(e_B, v) \geqq 0, \quad \forall v \in \mathscr{F}, \quad \tilde{v} \geqq 0 \quad \text{q.e.} \quad \text{on } B.$$

We call e_B the (1—) *equilibrium potential* of B. In view of (3.3.3) and Lemma 3.3.1, the equilibrium potential e_B is a 1-potential of a unique measure $\nu_B \in S_0$ supported by $\bar{B}$. ν_B is called the *equilibrium measure* of B. When A is open, we can see from Lemma 3.1.4 that e_A defined above is identical with the equilibrium potential already introduced at the beginning of §3.1. When A is a relatively compact open set, we also have

$$(3.3.4)\quad \nu_A(\bar{A}) = \mathrm{Cap}\,(A)$$

because, by taking $w \in \mathscr{F} \cap C_0(X)$ such that $w(x) = 1$, $\forall x \in \bar{A}$, we see with the aid of Lemma 3.1.1 (iv) that $\mathrm{Cap}(A) = \mathscr{E}_1(e_A, w) = \int_{\bar{A}} w(x) \nu_A(dx) = \nu_A(\bar{A})$. The following theorem extends relation (3.1.4).

Theorem 3.3.1. Let B be a Borel set of finite capacity. Then

$$(3.3.5)\quad \mathrm{Cap}\,(B) = \mathscr{E}_1(e_B, e_B) = \inf_{u \in \mathscr{L}_B} \mathscr{E}_1(u, u).$$

Proof. The second equality is clear from the definition. The first one appears in (3.1.4) for B open. Let us first prove this for a compact set K.

We can choose a decreasing sequence $\{A_n\}$ of open sets such that $\bigcap_{n=1}^{\infty} \bar{A}_n = K$ and Cap $(A_n) \downarrow$ Cap(K). Denote by e_n and ν_n the equilibrium potential and the equilibrium measure of A_n. Since $\mathscr{E}_1(e_n - e_m, e_n - e_m) = \text{Cap}\,(A_n) - \text{Cap}\,(A_m)$, $n > m$, we see that e_n is $\mathscr{E}_1$-convergent to some element $e_0 \in \mathscr{F}$. In order to get the equality Cap $(K) = \mathscr{E}_1(e_K, e_K)$, it suffices to show

(3.3.6) $\quad e_0 = e_K.$

Since $\tilde{e}_n = 1$ q.e. on A_n, e_0 satisfies (3.3.2) with $B = K$. On the other hand, Supp $[\nu_n] \subset \bar{A}_n$ and $\nu_n(\bar{A}_n) = \text{Cap}\,(A_n) \leq \text{Cap}\,(A_1)$ by (3.3.4). Hence a subsequence of ν_n vaguely converges to a positive Radon measure ν_0 concentrated on K. From $\mathscr{E}_1(e_n, v) = \int_{\bar{A}_1} v(x)\nu_n(dx)$ follows $\mathscr{E}_1(e_0, v) = \int_K v(x)\nu_0(dx)$, which, combined with Lemma 3.3.1, means that e_0 satisfies (3.3.3) with $B = K$. (3.3.6) is proven.

Consider next a Borel set B with Cap $(B) < \infty$. On account of the capacitability (3.1.6), we can find an increasing sequence $\{K_n\}$ of compact sets and a decreasing sequence $\{B_n\}$ of open sets such that

(3.3.7) $\quad K_n \subset B \subset A_n, \lim_{n\to\infty} \text{Cap}\,(K_n) = \text{Cap}\,(B) = \lim_{n\to\infty} \text{Cap}\,(A_n).$

By an argument similar to the preceding paragraph, we conclude that both e_{K_n} and e_{A_n} are $\mathscr{E}_1$-convergent to the same element $e_0 \in \mathscr{F}$. Since the property (3.3.2) (resp. (3.3.3)) of e_0 follows from the same property of e_{A_n} (resp. e_{K_n}), e_0 must coincide with e_B. So (3.3.5) is proven. Q.E.D.

Problem 3.3.1. Let $\mathscr{D}$ be a core of $\mathscr{E}$ satisfying the condition in Problem 3.2.1. Show that each condition of Lemma 3.3.1 is also equivalent to the following weaker one when the set F is compact:

(iv) $\mathscr{E}_\alpha(u, v) \geq 0, \quad \forall v \in \mathscr{D}, \quad v \geq 0 \;$ on $\; F.$

Problem 3.3.2. Let $\mathscr{D}$ be the same as above and K be a compact set. Show that Cap$(K) = \inf_{u \in \mathscr{D}_k} \mathscr{E}_1(u, u)$ where $\mathscr{D}_k = \{u \in \mathscr{D}: u(x) \geq 1 \quad \forall x \in K\}$.

(Hint: Prove that the limit of a minimizing sequence satisfies (3.3.2) and (3.3.3) by the help of Problem 3.3.1).

Problem 3.3.3. (Gauss's quadratic form). For a compact set K, Cap(K) equals the total mass of the equilibrium measure ν_K. Show that ν_k is the unique element in the class $\{\mu \in S_0\colon \text{Supp}[\mu] \subset K\}$ minimizing the quadratic form $\mathbf{G}(\mu) = \mathscr{E}_1(\mu) - 2\mu(K)$.

(Hint: Use Theorem 3.2.2 to get $\mathbf{G}(\mu) = \mathbf{G}(\nu_K) + \mathscr{E}_1(U_1\mu - U_1\nu_K, U_1\mu - U_1\nu_K)$).

Theorem 3.3.2 The following conditions are equivalent for a Borel set $B \subset X$:
(i) $\text{Cap}\,(B) = 0$
(ii) $\mu(B) = 0, \ {}^\forall\mu \in S_0$
(iii) $\mu(B) = 0, \ {}^\forall\mu \in S_{00}$
where $S_{00} = \{\mu \in S_0 : \mu(X) = 1, \|U_1\mu\|_\infty < \infty\}$.

Proof. If a Borel set B has positive capacity, Cap $(K) > 0$ for a certain compact set $K \subset B$ by (3.1.6). Then the equilibrium measure ν_K ($\in S_0$) has positive measure on B: $\nu_K(B) = \nu_K(K) = \text{Cap}\,(K)$ by (3.3.5) and (3.2.8). Combining this with Lemma 3.2.2, we know the equivalence of (i) and (ii).

Assume that a Borel set B satisfies (iii). Let us derive (ii) by utilizing the maximum principle Lemma 3.2.3. For any non-vanishing $\mu \in S_0$, put $\Gamma_n = \{x \in X : \widetilde{U_1\mu}(x) \leq n\}$. Choose compact sets K_n increasing to X and put $\mu_n(E) = \mu(E \cap \Gamma_n \cap K_n)/\mu(\Gamma_n \cap K_n)$. Since $\widetilde{U_1\mu_n} \leq n/\mu(\Gamma_n \cap K_n)$ q.e. on Γ_n, $\|U_1\mu_n\|_\infty$ is bounded by the same constant in view of Lemma 3.2.3. Hence $\mu_n \in S_{00}$ and $\mu(B) = \lim_{n\to\infty} \mu(\Gamma_n \cap K_n)\mu_n(B) = 0$. Q.E.D.

We now give a brief comment on reduced functions. The next simple lemma is useful in this connection.

Lemma 3.3.2. Let u_1 and u_2 be α-excessive functions in $L^2(X;m)$, $\alpha > 0$. If $u_1 \leq u_2$ m-a.e. and $u_2 \in \mathscr{F}$, then $u_1 \in \mathscr{F}$ and $\mathscr{E}_\alpha(u_1, u_1) \leq \mathscr{E}_\alpha(u_2, u_2)$.

Proof. $(u_1 - e^{-\alpha t}T_tu_1, u_1) \leq (u_1 - e^{-\alpha t}T_tu_1, u_2) = (u_1, u_2 - e^{-\alpha t}T_tu_2) \leq (u_2, u_2 - e^{-\alpha t}T_tu_2)$. Hence, $\lim_{t\downarrow 0} \mathscr{E}^{(t)}(u_1, u_1) + \alpha(u_1, u_1) \leq \mathscr{E}_\alpha(u_2, u_2) < \infty$ and the conclusion of the lemma follows. Q.E.D.

Consider any α-excessive function $f \in \mathscr{F}$ and any set $B \subset X$. We put

(3.3.8) $\quad \mathscr{L}_{f,B} = \{w \in \mathscr{F} : \tilde{w} \geq \tilde{f} \quad \text{q.e.} \quad \text{on } B\}$

In the same way as in the paragraph below (3.3.1), we see that $\mathscr{L}_{f,B}$ admits a unique element, say $f_B \in \mathscr{L}_{f,B}$, minimizing $\mathscr{E}_\alpha(w, w)$ on $\mathscr{L}_{f,B}$, $\alpha > 0$ being fixed. f_B is called the *α-reduced function* of f on B. Clearly

(3.3.9) $\quad \mathscr{E}_\alpha(f_B, v) \geq 0, \ {}^\forall v \in \mathscr{F}, \ \tilde{v} \geq 0 \quad \text{q.e.} \quad \text{on } B.$

On account of Theorem 3.2.1 and Lemma 3.3.1, f_B is α-excessive again and $f_B = U_\alpha \nu$ with a measure $\nu \in S_0$ supported by $\bar{B}$. Since $u = f_B \wedge f$ is α-excessive and $u \leqq f_B$ m-a.e., $\mathscr{E}_\alpha(u, u) \leq \mathscr{E}_\alpha(f_B, f_B)$ by Lemma 3.3.2. But this means $u = f_B$ because $u \in \mathscr{L}_{f,B}$. Hence

(3.3.10) $f_B \leqq f$ m-a.e.

(3.3.11) $\tilde{f}_B = \tilde{f}$ q.e. on B.

We have the following analogue of Lemma 3.1.1.

Lemma 3.3.3. The α-reduced function f_B on B of an α-excessive function $f \in \mathscr{F}$ is a unique element of $\mathscr{F}$ satisfying (3.3.9) and (3.3.11).

Given f and B as above, let
$\mathscr{U}_{f,B} = \{u\colon 0 \leqq u \leqq f$ m-a.e., u is α-excessive and $\tilde{u} = \tilde{f}$ q.e. on $B\}$.
Note that the first two conditions for $u \in \mathscr{U}_{f,B}$ imply that $u \in \mathscr{F}$ and thus the third condition for u makes sense.

Theorem 3.3.3. Let f and B be as above.
(i) The α-reduced function f_B of f on B is the minimum element of $\mathscr{U}_{f,B}$, i.e., $f_B \in \mathscr{U}_{f,B}$, $f_B \leqq u$, m-a.e., $\forall u \in \mathscr{U}_{f,B}$.
(ii) Given an m-measurable function u on X such that $0 \leqq u \leqq f_B$ m-a.e., u is α-excessive and $\tilde{u} = \tilde{f}$ q.e. on B, then $u = f_B$ m-a.e.
(iii) Let B be a Borel set of finite capacity and e_B be its equilibrium potential. Given an m-integrable function u on X such that $0 \leqq u \leqq e_B$ m-a.e., u is 1-excessive and $\tilde{u} = 1$ q.e. on B, then $u = e_B$ m-a.e.

Proof. (i) We have shown that $f_B \in \mathscr{U}_{f,B}$. For any $u \in \mathscr{U}_{f,B}$, let $u' = f_B \wedge u$. u' is then α-excessive and $u' \leqq f_B$. By Lemma 3.3.2 and the fact that $u' \in \mathscr{L}_{f,B}$, we get $f_B = u' \leqq u$ m-a.e.
(ii) If u satisfies the stated properties, then $u \in \mathscr{U}_{f,B}$ and consequently $f_B \leqq u$ by (i). In view of the first property of u, we get $u = f_B$.
(iii) Lemma 3.3.3 particularly means that $(e_B)_B = e_B$. If u satisfies the stated properties, $e_B = (e_B)_B \leqq u$ on account of (i) and (3.3.2). Combining this with the first property of u, we get $u = e_B$. Q.E.D.

The operation of taking the reduced functions induces a transformation on S_0—the so called "sweeping out". Given any $\mu \in S_0$, the α-reduced function $(U_\alpha\mu)_B$ of $U_\alpha\mu$ on B is again an α-potential of a unique measure, say μ_B, in S_0 supported by $\bar{B}$. We have already observed this right after (3.3.9). We call μ_B the *α-sweeping out* of μ on B. Thus

(3.3.12) $U_\alpha \mu_B = (U_\alpha \mu)_B,\ \mu \in S_0.$

It is important to look at the operation of taking the reduced functions from another point of view. Let us introduce the space

(3.3.13) $\mathscr{F}_{X-B} = \{u \in \mathscr{F} : \tilde{u} = 0 \quad \text{q.e.} \quad \text{on } B\}$

for a Borel set B. $\mathscr{F}_{X-B}$ is a closed subspace of the Hilbert space $(\mathscr{F}, \mathscr{E}_\alpha)$ for each $\alpha > 0$. Denote by $\mathscr{H}^B_\alpha$ its orthogonal complement:

(3.3.14) $\mathscr{F} = \mathscr{F}_{X-B} \oplus \mathscr{H}^B_\alpha.$

In view of (3.3.9) and (3.3.11), $f = (f - f_B) + f_B$ reperesents the corresponding orthogonal decomposition of the α-excessive function $f \in \mathscr{F}$. Thus the α-reduced function f_B on the set B is nothing more than the projection of f on the space $\mathscr{H}^B_\alpha$.

In connection with the space $\mathscr{H}^B_\alpha$, we finally prove a theorem stating in a certain sense that the "spectral synthesis" is possible for th Dirichlet space.[1] We say that an open set G is an *α-regular set* of $u \in \mathscr{F}$ if for any $v \in \mathscr{F} \cap C_0(X)$ with Supp $[v] \subset G$,

(3.3.15) $\mathscr{E}_\alpha(u, v) = 0.$

The union $G_1 \cup G_2$ of two α-regular open sets G_1 and G_2 of u is again α-regular for u. In fact, for any $v \in \mathscr{F} \cap C_0(X)$ with Supp $[v] \subset G_1 \cup G_2$, take a relatively compact open set E such that Supp $[v] - G_2 \subset E \subset \bar{E} \subset G_1$ and choose a function $w \in \mathscr{F} \cap C_0(X)$ such that $w = 1$ on E and Supp $[w] \subset G_1$ (Theorem 1.4.2 and Lemma 1.4.2). Then $v = v_1 + v_2$ with $v_1 = vw$ and $v_2 = v - vw$. Since $v_i \in \mathscr{F} \cap C_0(X)$ and Supp $[v_i] \subset G_i$, $i = 1, 2$, (3.3.15) is valid for the present v.

This observation enables us to define the *α-singular set* or the *α-spectrum* $\sigma_\alpha(u)$ of $u \in \mathscr{F}$ as the complement of the largest α-regular open set of u. It is then clear that

(3.3.16) $\sigma_\alpha(U_\alpha \mu) = \text{Supp}\,[\mu], \quad \mu \in S_0.$

1) Cf. J. P. Kahane, Quotients de fonctions définies-négatives, Séminaire Bourbaki, 19 (1966/67).

Lemma 3.3.4. Let G be an open set and W_α^G be the $\mathscr{E}_1$-closure of the space of functions $u \in \mathscr{F}$ with $\sigma_\alpha(u) \subset G$. Then $W_\alpha^G = \mathscr{H}_\alpha^G$.

Proof. If $u \in \mathscr{F}$ is $\mathscr{E}_\alpha$-orthogonal to W_α^G, then $(u, f) = \mathscr{E}_\alpha(u, G_\alpha f) = 0$ for any bounded function f of $L^2(X; m)$ with Supp $[f] \subset G$ because of (3.3.16). Hence, $u \in \mathscr{F}_{X-G}$ by Lemma 3.1.4.

Conversely, let us show that $v \in \mathscr{F}_{X-G}$ is $\mathscr{E}_\alpha$-orthogonal to any $u \in \mathscr{F}$ with $\sigma_\alpha(u) \subset G$. This is evident when $v \in \mathscr{F} \cap C_0(X)$. Suppose that $v \in \mathscr{F}_{X-G}$ is non-negative bounded and of compact support. Take $f \in \mathscr{F} \cap C_0(X)$ such that $f \geq v$ and Supp$[f] \cap \sigma_\alpha(u) = \phi$. Choose a sequence of non-negative functions $v_n \in \mathscr{F} \cap C_0(X)$ $\mathscr{E}_1$-convergent to v. Then $f \wedge v_n = 1/2\,(f + v_n) - 1/2\,|f - v_n|$ is $\mathscr{E}_1$-convergent to v because $|f - v_n|$ is $\mathscr{E}_1$-convergent to $f - v$ by virtue of Theorem 1.4.2, (v). Hence, $\mathscr{E}_\alpha(u, v) = \lim_{n\to\infty} \mathscr{E}_\alpha(u, f \wedge v_n) = 0$. For a general non-negative $v \in \mathscr{F}_{X-G}$, we take $v_n \in \mathscr{F} \cap C_0(X)$ as above. Then $v \wedge v_n$ is $\mathscr{E}_1$-convergent to v and consequently $\mathscr{E}_\alpha(u, v) = 0$. Q.E.D.

Theorem 3.3.4. Let F be a closed set and W_α^F be the space of functions $u \in \mathscr{F}$ with $\sigma_\alpha(u) \subset F$. Then

(3.3.17) $$W_\alpha^F = \mathscr{H}_\alpha^F.$$

In particular, any function $u \in \mathscr{F}$ can be approximated in the $\mathscr{E}_1$-metric by finite linear combinations of α-potentials of measures in S_0 supported by $\sigma_\alpha(u)$.

Proof. The inclusion $W_\alpha^F \supset \mathscr{H}_\alpha^F$ is trivial. To get the converse, observe that

(3.3.18) $$W_\alpha^F = \bigcap_{n=1}^\infty W_\alpha^{G_n}$$

where $\{G_n\}$ is a sequence of open sets such that $G_n \supset \bar{G}_{n+1}$, $\bigcap_{n=1}^\infty G_n = F$. Indeed, we have $\sigma_\alpha(u) \subset \bar{G}_n$ for $u \in W_\alpha^{G_n}$, $n = 1, 2, \cdots$.

(3.3.18) and Lemma 3.3.4 mean that the $\mathscr{E}_\alpha$-projection $P_{W_\alpha^F} u$ on W_α^F of an α-excessive function $u \in \mathscr{F}$ is the $\mathscr{E}_1$-convergent limit of its projections on $\mathscr{H}_\alpha^{G_n}$, which are the α-reduced functions u_{G_n} according to the preceding remark. Therefore, $\mathbf{P}_{W_\alpha^F} u$ is α-excessive again and an α-potential of a measure in S_0 supported by F in view of (3.3.16). In particular, $\mathbf{P}_{W_\alpha^F} u \in \mathscr{H}_\alpha^F$ by virtue of Lemma 3.3.1.

Now take, for any $u \in W_\alpha^F$, a sequence $u_n = G_\alpha f_n$, $f_n \in L^2(X; m)$, which is $\mathscr{E}_1$-convergent to u. Since u_n is difference of α-excessive functions and $\mathbf{P}_{W_\alpha^F} u_n$ is $\mathscr{E}_1$-convergent to u, we get the inclusion $W_\alpha^F \subset \mathscr{H}_\alpha^F$ together with the latter statement in Theorem 3.3.4. Q.E.D.

Problem 3.3.4. Let $\mathscr{D}$ be a core of $\mathscr{E}$ satisfying the condition in Problem 3.2.1. Furthermore, assume that for any compact set K and a relatively compact open set $G \supset K$, $\mathscr{D}$ contains an element u such that $u(x) = 1$, $x \in K$, $u(x) = 0$, $x \in X - G$.

Show then that an open set G is α-regular for $u \in \mathscr{F}$ if and only if (3.3.15) holds for any $v \in \mathscr{D}$ with $\mathrm{Supp}[v] \subset G$.
(Hint: See the proof of Theorem 2.1.2.).

Except for Lemma 3.3.2 and Theorem 3.3.3, each assertion in this section admits its obvious 0-order version when $(\mathscr{F}, \mathscr{E})$ is transient. See the same remarks made at the end of §3.1 and §3.2. The proof of Lemma 3.3.2 uses the fact that $\mathscr{F} \subset L^2(X; m)$ and hence the extension to $\mathscr{F}_e$ is not obvious.

Example 3.3.1. (Capacities in the classical potential theory).
Let $(\mathscr{F}_e, \mathscr{E})$ be the extended Dirichlet space of the transient symmetric stable semi-group of index α $(0 < \alpha \leqq 2, \alpha < n)$. Consider the associated (0-order) capacity $\mathrm{Cap}_{(0)}$ introduced at the end of §3.1. Then Example 3.2.1 and the 0-order version of Problem 3.3.3 imply the following: for any compact $K \subset R^n$,

$$\mathrm{Cap}_{(0)}(K) = \mu_K(K) \tag{3.3.19}$$

where μ_K is the unique element in the class

$$\{\mu: \text{positive Radon measure on } R^n,\ \mathrm{Supp}[\mu] \subset K,\ \textstyle\int_{R^n}\int_{R^n} R^{(\alpha)}(x-y)\mu(dx)\mu(dy) < \infty\}$$

minimizing the Gauss quadratic form $\mathbf{G}(\mu) = \int_{R^n}\int_{R^n} R^{(\alpha)}(x-y)\mu(dx)\mu(dy) - 2\mu(K)$. This is exactly one of the equivalent definitions of capacity in classical potential theory.

In particular, the Newtonian capacity $\mathrm{C}(K)$ of a compact set $K \subset R^n$, $n \geqq 3$, corresponds to the case where $\alpha = 2$, $c = 1/2$. We see from (3.3.19) and the 0-order version of Problem 3.3.2 that $\mathrm{C}(K) = \inf 1/2\, \mathbf{D}(u, u)$, where the infimum on the right hand side is taken over all $u \in C_0^\infty(R^n)$ such that $u(x) \geqq 1$, $x \in K$.

Example 3.3.2. (Spectral synthesis in the Sobolev space).
Consider the Dirichlet form $(\mathrm{H}^1(R^n), 1/2\, \mathbf{D})$ on $L^2(R^n)$ $(n \geqq 1)$. Then for any $u \in H^1(R^n)$,

$$\sigma_\alpha(u) = \mathrm{Supp}\left[\left(\alpha - \frac{1}{2}\Delta\right)u\right], \quad \alpha \geqq 0, \tag{3.3.20}$$

where $(\alpha - 1/2\,\Delta)u$ is understood in the sense of distribution. This follows from $\mathscr{E}_\alpha(u, v) = \langle(\alpha - 1/2\Delta)u, v\rangle$, $v \in C_0^\infty(R^n)$, and Problem 3.3.4.

For a domain $D \subset R^n$, we may regard the space $H_0^1(D)$ of Example 1.2.3 as a closed subspace of $H^1(R^n)$. As a consequence of the theorem of the spectral synthesis (Theorem 3.3.4), we then have

$$(3.3.21)\quad H_0^1(D) = \mathscr{F}_D$$

where $\mathscr{F}_D = \{u \in H^1(R^n),\ \tilde{u} = 0 \text{ q.e. on } R^n - D\}$. In fact, $H_0^1(D)$ is the $\mathscr{E}_\alpha$-orthogonal complement of the space $W_\alpha^{R^n - D}$ in view of Problem 3.3.4.

When $n \geqq 3$, the relation (3.3.20) holds for $\alpha = 0$ and for any u in the extended Dirichlet space $\mathscr{F}_e$. The 0-order version of the spectral synthesis states that any $u \in \mathscr{F}_e$ can be approximated in the Dirichlet norm by Newtonian potentials of signed Radon measures supported by $\sigma_0(u)$.

Example 3.3.3. (one-dimensional case). Let $I = (r_1, r_2)$ be a one-dimensional interval and let m and k be positive Radon measures on I with Supp $[m] = \mathrm{I}$. We consider the regular Dirichlet space $(\mathscr{F}^0, \mathscr{E})$ on $L^2(I; m)$ introduced in Example 1.2.2 by (1.2.7) and (1.2.9).

We saw in Example 1.2.2 that any $\mathscr{E}_1$-Cauchy sequence $u_n \in \mathscr{F}^0$ is uniformly convergent on each compact subinterval of I. Therefore there exists for each $y \in I$ a unique $g_1^0(\cdot, y) \in \mathscr{F}^0$ such that

$$(3.3.22)\quad \mathscr{E}_1(g_1^0(\cdot, y), v) = v(y), \quad v \in \mathscr{F}^0.$$

$g_1^0(x,y)$ is the so-called *reproducing kernel*. We note that $0 < g_1^0(y,y) < \infty$ for any $y \in I$. If $g_1^0(y, y) = \mathscr{E}_1(g_1^0(\cdot, y), g_1^0(\cdot, y))$ were non-positive, then $g_1^0(x,y) = 0$, $^\forall x \in I$, and $v = 0$, $^\forall v \in \mathscr{F}^0$, by (3.3.22), arriving at a contradiction.

Let us show that

$$(3.3.23)\quad \mathrm{Cap}(\{y\}) = 1/g_1^0(y,y) \qquad (> 0).$$

Take the 1-equilibrium potential $p_n (\in \mathscr{F}^0)$ of the open interval $J_n = (y - 1/n, y + 1/n)$. Just as in the proof of Theorem 3.3.1, p_n converges to a function $p \in \mathscr{F}^0$ with metric $\mathscr{E}_1$ and, consequently, pointwise. p is characterized by the property

$$(3.3.24)\quad p \in \mathscr{F}^0,\ p(y) = 1, \quad \mathscr{E}_1(p, v) \geqq 0 \quad {}^\forall v \in \mathscr{F}^0, \quad v(y) \geqq 0.$$

In fact, $p(y) = 1$ follows from the fact that the continuous function p_n equals 1 m-a.e. on J_n. If $v \in \mathscr{F}^0$ is non-negative at y, then $v_\delta(x) = v(x) + \delta g_1^0(x,y)$ is positive on a neighbourhood of y for each $\delta > 0$. Hence, $\mathscr{E}_1(p, v_\delta) = \mathscr{E}_1(p, v) + \delta\mathscr{E}_1(p, g_1^0(\cdot, y)) = \lim_{n\to\infty} \mathscr{E}_1(p_n, v_\delta) \geqq 0$. By letting $\delta \downarrow 0$, we get the second property in (3.3.24). A comparison of (3.3.22) with (3.3.24) gives

$$(3.3.25)\quad p(x) = g_1^0(x, y)/g_1^0(y,y).$$

(3.3.23) follows from this and $\mathrm{Cap}(\{y\}) = \mathscr{E}_1(p,p)$.

Thus every non-empty set has a positive capacity in the present example. In particular, (3.3.24) characterizes the 1-equilibrium potential p of the one point set $\{y\}$. Using this again we can readily derive another expression of p:

$$(3.3.26) \quad p(x) = \begin{cases} u_1(x)/u_1(y) & \text{if} \quad x \leqq y, \\ u_2(x)/u_2(y) & \text{if} \quad x \geqq y, \end{cases}$$

where u_1(resp. u_2) is a strictly increasing (resp. decreasing) positive 1-harmonic function on I such that $u_1(r_1) = 0$ (resp. $u_2(r_2) = 0$) if r_1 (resp. r_2) is a regular boundary (see Example 1.2.2).

(3.3.25), (3.3.26), and the symmetry of $g_1^0(x,y)$ then lead us to

$$(3.3.27) \quad g_1^0(x, y) = \begin{cases} Cu_1(x)u_2(y) & \text{if} \quad x \leqq y, \\ Cu_2(x)u_1(y) & \text{if} \quad x \geqq y, \end{cases}$$

for some constant $C > 0$. This expression along with (3.3.22) enables us to obtain

$$(3.3.28) \quad \begin{cases} R_1 f(x) = \int_{r1}^{r2} g_1^0(x, y) f(y) dy, \\ U_1 \mu(x) = \int_{r1}^{r2} g_1^0(x, y) \mu(dy), \end{cases}$$

for a non-negative Borel $f \in L^2(I;m)$ and $\mu \in S_0$.

Let $\{F_n\}$ be an increasing sequence of closed subsets of I such that $\mathrm{Cap}(I - F_n) \to 0$, $n \to \infty$. Then any compact subinterval of I is contained in some F_n because $\mathrm{Cap}(\{y\})$ is a positive continuous function on I by virtue of (3.3.23) and (3.3.27). In particular, the quasi-continuity reduces to the ordinary continuity.

Part II

Symmertic Markov Processes

CHAPTER 4

Analysis in Symmetric Markov Processes

A Hunt process is a special Markov process that possesses useful properties such as right continuity of sample paths, quasi-left continuity, and strong Markov property (§4.1). In sections 4.3, 4.4, and 4.5, we assume that we are given a symmetric Hunt process whose Dirichlet form is regular. Such an assumption imposes no restriction upon a regular Dirichlet form however, for, as we shall see in §6.2, any regular Dirichlet form admits a Hunt process.

In §4.3 and §4.4, the potential theoretic notions and relations developed in the preceding chapter are interpreted in terms of the Hunt process. In particular, the part of the Hunt process on an open set (the absorbing barrier process) is shown to correspond to the part of the Dirichlet form on the set. The regularity of the latter is an immediate consequence of spectral synthesis.

The local property of the Dirichlet form is a necessary and sufficient condition for the Hunt process to possess continuous sample paths with probability 1 (§4.5). §4.5 also justifies our saying that the measures k and J appearing in the Beurling-Deny formula are the killing and jumping measures respectively.

§4.1 Elements in the Hunt Process

First of all, we give a definition of the Hunt process in a manner slightly different from (but equivalent to) R. M. Blumenthal and R. K. Getoor [14] for later convenience.

Given a measurable space $(S, \mathscr{B})$, the completion of the σ-field $\mathscr{B}$ with respect to a probability measure μ is denoted by $\mathscr{B}^{\mu}$. An element B of $\mathscr{B}^* = \bigcap_{\mu \in \mathscr{P}(S)} \mathscr{B}^{\mu}$ is called a *universally measurable set*, where $\mathscr{P}(S)$ denotes the family of all probability measures on S. Given a sub-σ-field $\mathscr{C}$, we say that $\mathscr{C}^{\mu}$ is the *completion of* $\mathscr{C}$ *in* $\mathscr{B}^{\mu}$ *with respect to* μ if $\mathscr{C}^{\mu} = \{A \subset S : A \varDelta B \subset N, \exists B \in \mathscr{C}, \exists N \in \mathscr{B}, \mu(N) = 0\} = \sigma\{\mathscr{C}, \mathscr{N}\}$. Here $\sigma\{\ldots\}$ denotes the smallest σ-field making things inside the braces measurable and $\mathscr{N} = \{N \in \mathscr{B}^{\mu}\colon \mu(N) = 0\}$.

Let $(\Omega, \mathcal{M}, \{X_t\}_{t\in T}, P)$ be a *stochastic process* with *state space* $(S, \mathcal{B})$ and *time parameter* $T \subset [0, \infty]$, i.e., $(S, \mathcal{B})$ is a measurable space, $(\Omega, \mathcal{M}, P)$ is a probability space, and X_t is a measurable map from Ω to S for each $t \in T$. The last condition of the measurability is explicitly indicated by $X_t \in \mathcal{M}/\mathcal{B}$. We set

$$(4.1.1) \quad \begin{cases} \mathcal{F}_\infty^0 = \sigma\{X_s, s \in T, s < \infty\} \\ \mathcal{F}_t^0 = \sigma\{X_s, s \in T, s \leqq t\}, t \in T, t < \infty. \end{cases}$$

Assume that $T = [0, \infty]$. We say that a family $\{\mathcal{M}_t\}_{t\geqq 0}$ of sub-σ-fields of $\mathcal{M}$ is *admissible* if $\mathcal{M}_t$ is increasing in t and $X_t \in \mathcal{M}_t/\mathcal{B}$ for each $t \geqq 0$. We may then call $\{\mathcal{F}_t^0\}_{t\geqq 0}$ the *minimum admissible family*. Given an admissible family $\{\mathcal{M}_t\}_{t\geqq 0}$, a $[0, \infty]$-valued function σ on Ω is called an $\{\mathcal{M}_t\}$*-stopping time* if $\{\sigma \leqq t\} \in \mathcal{M}_t$, $t \geqq 0$. We can then define a sub-σ-field $\mathcal{M}_\sigma$ by $\mathcal{M}_\sigma = \{\Lambda \in \mathcal{M}: \Lambda \cap \{\sigma \leqq t\} \in \mathcal{M}_t, \forall t \geqq 0\}$. An admissible family is called *right continuous* if $\mathcal{M}_t = \mathcal{M}_{t+}(= \bigcap_{t'>t} \mathcal{M}_{t'})$, $\forall t \geqq 0$.

Adjoining an extra point Δ to a measurable space $(S, \mathcal{B})$, we set $S_\Delta = S \cup \Delta$ and $\mathcal{B}_\Delta = \mathcal{B} \cup \{B \cup \Delta: B \in \mathcal{B}\}$. A quadruplet $\mathbf{M} = (\Omega, \mathcal{M}, \{X_t\}_{t\in[0,\infty]}, \{P_x\}_{x\in S_\Delta})$ is said to be a *Markov process on* $(S, \mathcal{B})$ *with time parameter* $[0, \infty]$ if the following four conditions are satisfied:

(**M.** 1) For each $x \in S_\Delta$, $(\Omega, \mathcal{M}, \{X_t\}_{t\in[0,\infty]}, P_x)$ is a stochastic process with state space $(S_\Delta, \mathcal{B}_\Delta)$ and parameter set $[0, \infty]$,

(**M.** 2) $P_x(X_t \in E)$ is $\mathcal{B}$-measurable in $x \in S$ for each $t \geqq 0$ and $E \in \mathcal{B}$,

(**M.** 3) There exists an admissible family $\{\mathcal{M}_t\}_{t\geqq 0}$ such that

$$(4.1.2) \quad P_x(X_{t+s} \in E \mid \mathcal{M}_t) = P_{X_t}(X_s \in E),\ P_x\text{-a.s.}$$

for any $x \in S$, $t, s \geqq 0$ and $E \in \mathcal{B}$.

(**M.** 4) $P_\Delta(X_t = \Delta) = 1$, $\forall t \geqq 0$.

If an additional condition

(**M.** 5) $P_x(X_0 = x) = 1$, $x \in S$

is satisfied, then the Markov process $\mathbf{M}$ is called *normal*. The condition (4.1.2) is called the *Markov property with respect to the admissible family* $\{\mathcal{M}_t\}$. As an immediate consequence of this property the function $p_t(x, E)$ defined by $p_t(x, E) = P_x(X_t \in E)$, $x \in S$, $t \geqq 0$, $E \in \mathcal{B}$, is a Markovian transition function in the sense of §1.4. We call this the *transition function of the Markov process* $\mathbf{M}$.

Lemma 4.1.1. Let $\mathbf{M}$ be a Markov process.

(i) $\mathbf{M}$ posesses the Markov property with respect to the minimum admissible family $\{\mathcal{F}_t^0\}_{t\geqq 0}$.

(ii) $P_x(\Lambda)$ is $\mathscr{B}_\Delta$-measurable in $x \in S_\Delta$ for any $\Lambda \in \mathscr{F}^0_\infty$.

Given a Markov process **M** as above, $P_\mu(\Lambda) = \int_{S_\Delta} P_x(\Lambda)\mu(dx)$, $\mu \in \mathscr{P}(S_\Delta)$, $\Lambda \in \mathscr{F}^0_\infty$, defines a probability measure P_μ on $(\Omega, \mathscr{F}^0_\infty)$. We then denote by $\mathscr{F}^\mu_\infty$ (resp. $\mathscr{F}^\mu_t$) the completion of $\mathscr{F}^0_\infty$ (resp. completion of $\mathscr{F}^0_t$ in $\mathscr{F}^\mu_\infty$) with respect to P_μ. We also set $\mathscr{F}_\infty = \bigcap_{\mu \in \mathscr{P}(S_\Delta)} \mathscr{F}^\mu_\infty$, $\mathscr{F}_t = \bigcap_{\mu \in \mathscr{P}(S_\Delta)} \mathscr{F}^\mu_t$. Obviously, $\{\mathscr{F}^0_{t+}\}$, $\{\mathscr{F}^\mu_t\}$ $\{\mathscr{F}_t\}$ are admissible families. We may call $\{\mathscr{F}_t\}_{t \geq 0}$ the *minimum completed admissible family*.

Lemma 4.1.2. ([14; I, (8.12)]). If **M** has the Markov property with respect to $\{\mathscr{F}^0_{t+}\}$, then $\{\mathscr{F}^\mu_t\}$ and $\{\mathscr{F}_t\}$ are right continuous.

Lemma 4.1.3. (E. B. Dynkin [30; Lemma 3.4], P. A. Meyer [66; Chap. XIII, T20]). Let $\mu \in \mathscr{P}(S_\Delta)$.
(i) For any $\{\mathscr{F}^\mu_t\}$-stopping time σ, there exists an $\{\mathscr{F}^0_{t+}\}$-stopping time σ' such that $P_\mu\,(\sigma \neq \sigma') = 0$.
(ii) For any $\Lambda \in \mathscr{F}^\mu_\sigma$, there exists $\Lambda' \in \mathscr{F}^0_{\sigma'+}$ such that $P_\mu(\Lambda \Delta \Lambda') = 0$, where $\mathscr{F}^0_{\sigma'+} = \{\Lambda \in \mathscr{F}^0_\infty : \Lambda \cap \{\sigma' \leqq t\} \in \mathscr{F}^0_{t+}, t \geqq 0\}$.

We are now in a position to give the definition of the Hunt process. Let X be a locally compact separable Hausdorff space and X_Δ be its one-point compactification. When X is already compact, Δ is adjoined as an isolated point. $\mathscr{B}(X)$ denotes the family of all Borel subsets of X. We fix a set $S \in \mathscr{B}(X)$ and regard $S_\Delta = S \cup \Delta$ as a topological subspace of X_Δ. We then define $\mathscr{B} = \mathscr{B}(S)$ as the family of all Borel subsets of S and $\mathscr{B}_\Delta = \mathscr{B} \cup \{B \cup \Delta : B \in \mathscr{B}(S)\}$.

Let us consider a normal Markov process $\mathbf{M} = (\Omega, \mathscr{M}, X_t, P_x)$ on $(S, \mathscr{B})$ and assume the following additional conditions concerning the pair (Ω, X_t):
(M. 6) (i) $X_\infty(\omega) = \Delta$, $\forall \omega \in \Omega$, (ii) $X_t(\omega) = \Delta$, $\forall t \geqq \zeta(\omega)$, where $\zeta(\omega) = \inf\,\{t \geqq 0 : X_t(\omega) = \Delta\}$, (iii) for each $t \in [0, \infty]$, there exists a map θ_t from Ω to Ω such that $X_s \circ \theta_t = X_{s+t}$, $s \geqq 0$, (iv) for each $\omega \in \Omega$, the sample path $t \rightsquigarrow X_t(\omega)$ is right continuous on $[0, \infty)$ and has the left limit on $(0, \infty)$ (inside S_Δ).

The θ_t and ζ which occur in the preceding paragraph are called the *translation operator* and the *life time* of **M** respectively.

For an admissible family $\{\mathscr{M}_t\}$, we then say that **M** is *strong Markov with respect to* $\{\mathscr{M}_t\}$ if $\{\mathscr{M}_t\}$ is right continuous and

$$(4.1.3)\quad P_\mu(X_{\sigma+s} \in E / \mathscr{M}_\sigma) = P_{X_\sigma}(X_s \in E),\quad P_\mu\text{-a.s.},\quad \mu \in \mathscr{P}(S_\Delta),\quad E \in \mathscr{B}_\Delta,\ s \geqq 0$$

for any $\{\mathcal{M}_t\}$-stopping time σ.
Property (4.1.3) is equivalent to

$$(4.1.3)' \quad P_\mu(X_{\sigma+s} \in E, \Lambda) = E_\mu(P_{X_\sigma}(X_s \in E); \Lambda), \ {}^\forall\Lambda \in \mathcal{M}_\sigma,^{1)}$$

because $X_\sigma \in \mathcal{M}_\sigma/\mathcal{B}$ by virtue of the right continuity of $\{\mathcal{M}_t\}$ and the sample path ([65; IV, T49]). We say that $\mathbf{M}$ is *quasi-left continuous* (on $(0, \infty)$) if for any $\{\mathcal{M}_t\}$-stopping times σ_n increasing to σ

$$(4.1.4) \quad P_\mu(\lim_{n\to\infty} X_{\sigma_n} = X_\sigma, \sigma < \infty) = P_\mu(\sigma < \infty), \mu \in \mathcal{P}(S_\Delta).$$

A Markov process $\mathbf{M} = (\Omega, \mathcal{M}, X_t, P_x)$ on $(S, \mathcal{B})$ satisfying the condition (M. 6) is called a *Hunt process* if there exists an admissible family $\{\mathcal{M}_t\}$ such that $\mathbf{M}$ is strong Markov and quasi-left continuous with respect to $\{\mathcal{M}_t\}$.

Notice that the right continuity of $\{\mathcal{M}_t\}$ is implied in the above definition of the Hunt process. If the conditions of the regularity of the sample path (M. 6) (iv) and the quasi-left continuity (4.1.4) are weakened by shortening the time interval $(0, \infty)$ into $(0, \zeta(\omega))$, then the above definition of the Hunt process reduces to the definition of the *standard process.*

Theorem 4.1.1. The following conditions are equivalent for a normal Markov process $\mathbf{M} = (\Omega, \mathcal{M}, X_t, P_x)$ on $(S, \mathcal{B})$ satisfying (M. 6):
(i) $\mathbf{M}$ is a Hunt process,
(ii) $\mathbf{M}$ is strong Markov and quasi-left continuous with respect to the minimum completed admissible family $\{\mathcal{F}_t\}_{t\geq 0}$,
(iii) For each $\mu \in \mathcal{P}(S_\Delta)$, $\{\mathcal{F}_t^\mu\}$ is right continuous and the identities (4.1.3) and (4.1.4) hold for $\{\mathcal{F}_t^\mu\}$-stopping times and for the single measure P_μ.

Proof. Since the implication (iii) ⇒ (ii) ⇒ (i) is trivial, it suffices to prove (i) ⇒ (iii). Suppose that $\mathbf{M}$ is strong Markov and quasi-left continuous with respect to an admissible family $\{\mathcal{M}_t\}$. Since $\mathcal{F}_{t+}^0 \subset \mathcal{M}_{t+} = \mathcal{M}_t$, $\mathbf{M}$ is strong Markov with respect to $\{\mathcal{F}_{t+}^0\}$. By Lemma 4.1.2, $\{\mathcal{F}_t^\mu\}$ is then right continuous for each fixed $\mu \in \mathcal{P}(S_\Delta)$. Let σ be an $\{\mathcal{F}_t^\mu\}$-stopping time and $\Lambda \in \mathcal{F}_\sigma^\mu$. Choose $\{\mathcal{F}_{t+}^0\}$-stopping time σ' and $\Lambda' \in \mathcal{F}_{\sigma'+}^0$ according to Lemma 4.1.3. Then the identity (4.1.3)′ for P_u, σ and

1) E_μ denotes the expectation with respect to P_μ; i.e., $E_\mu(F) = \int_\Omega F(\omega)P_\mu(d\omega)$ for a μ-measurable function F on Ω. $E_\mu(F;\Lambda)$ stands for $E_\mu(F\cdot I_\Lambda)$, $\Lambda \in \mathcal{M}$.

Λ follows from the same one for P_μ, σ' and Λ' expressing the strong Markov property with respect to $\{\mathscr{F}^0_{t+}\}$. The identity (4.1.4) for $\{\mathscr{F}_t\}$-stopping times and P_μ can be proved in the same way. Q.E.D.

There is an obvious version of Theorem 4.1.1 for the standard process. We use it without further specification. In order to illustrate the usefulness of Theorem 4.1.1 and its version for the standard process, we now consider some rather trivial transformations of the Hunt process.

Let $\mathbf{M} = (\Omega, \mathscr{M}, X_t, P_x)$ be a Hunt process on $(S, \mathscr{B})$. We define three kinds of *hitting times* σ_B, $\dot{\sigma}_B$ and $\hat{\sigma}_B$ of a Borel set $B \in \mathscr{B}(X)$ by

$$(4.1.5)\quad \begin{cases} \sigma_B(\omega) = \inf\{t > 0 : X_t(\omega) \in B\} \\ \dot{\sigma}_B(\omega) = \inf\{t \geqq 0 : X_t(\omega) \in B\} \\ \hat{\sigma}_B(\omega) = \inf\{t > 0 : X_{t-}(\omega) \in B\}. \end{cases}$$

Each of them is defined to be ∞ if the set in the braces is empty. According to [65; IV, 53], each of them is an $\{\mathscr{F}_t\}$-stopping time.

We say that a subset $\tilde{S} \subset S$, $\tilde{S} \in \mathscr{B}(X)$, is $\mathbf{M}$-*invariant* (or $\mathbf{M}$-*closed*) if $P_x(\tilde{\Omega}) = 1$, $\forall x \in \tilde{S}$, where

$$(4.1.6)\quad \tilde{\Omega} = \{\omega \in \Omega : X_t(\omega) \in \tilde{S}_\Delta,\ X_{t-}(\omega) \in \tilde{S}_\Delta,\ \forall t \geqq 0\}.$$

$\tilde{\Omega}$ is an element of $\mathscr{F}_\infty$ because $\tilde{\Omega} = \{\dot{\sigma}_{S-\tilde{S}} = \infty,\ \hat{\sigma}_{S-\tilde{S}} = \infty\}$. For an $\mathbf{M}$-invariant set $\tilde{S}$, we put

$$(4.1.7)\quad \mathbf{M}|_{\tilde{S}} = (\tilde{\Omega}, \mathscr{M} \cap \tilde{\Omega}, \{X_t\}_{t\in[0,\infty]}, \{P_x\}_{x\in \tilde{S}_\Delta})$$

and call this the *restriction of the process* $\mathbf{M}$ *to* $\tilde{S}$.

Theorem 4.1.2. The restriction $\mathbf{M}|_{\tilde{S}}$ of the Hunt process $\mathbf{M}$ to an $\mathbf{M}$-invariant Borel subset $\tilde{S}$ is again a Hunt process.

Proof. Since the conditions (M.1), (M.2), (M.3), (M.4), (M.5) and (M.6) for $\mathbf{M}|_{\tilde{S}}$ are evident, it suffcies to show the condition (iii) of Theorem 4.1.1 for $\mathbf{M}|_{\tilde{S}}$. Take a $\mu \in \mathscr{P}(\tilde{S}_\Delta)$ and regard this as an element of $\mathscr{P}(S_\Delta)$ as well. We indicate the typical σ-fields of the process $\mathbf{M}|_{\tilde{S}}$ by adding $\sim$. It then follows from $\tilde{\mathscr{F}}^0_t = \mathscr{F}^0_t \cap \tilde{\Omega}$, $\tilde{\Omega} \in \mathscr{F}^\mu_\infty$ and $P_\mu(\tilde{\Omega}) = 1$ that $\tilde{\mathscr{F}}^\mu_t = \mathscr{F}^\mu_t \cap \tilde{\Omega} \subset \mathscr{F}^\mu_t$. Hence, the property of Theorem 4.1.1 (iii) for $\mathbf{M}$ implies the same property for $\mathbf{M}|_{\tilde{S}}$. Q.E.D.

Conversely, the Hunt process $\mathbf{M}$ on S can be extended to a larger state space $\hat{S}$ in a trivial manner as follows:

Theorem 4.1.3. Consider any set $\hat{S} \in \mathscr{B}(X)$ containing S. Then there exists a Hunt process $\hat{\mathbf{M}} = (\hat{\Omega}, \hat{\mathscr{M}}, \hat{X}_t, \hat{P}_x)$ on $(\hat{S}, \mathscr{B}(\hat{S}))$ such that
(i) S is $\hat{\mathbf{M}}$-invariant and $\hat{\mathbf{M}}|_S = \mathbf{M}$, where $\hat{\mathbf{M}}|_S$ denotes the restriction of $\hat{\mathbf{M}}$ to S,
(ii) each point $x \in \hat{S} - S$ is a trap with respect to $\hat{\mathbf{M}}$:
$\hat{P}_x(\hat{X}_t = x, {}^\forall t \geq 0) = 1.$

Proof. Set $\Omega_0 = \hat{S} - S$, $\mathscr{B}(\Omega_0) = \mathscr{B}(\hat{S} - S)$. Each point $x \in \hat{S} - S$ is also denoted by ω_x. Adjoining Ω_0 to Ω as an extra set, we put $\hat{\Omega} = \Omega \cup \Omega_0$. It then suffices to define other elements of $\hat{\mathbf{M}}$ by $\hat{\mathscr{M}} = \{\Lambda \cup \Lambda_0 : \Lambda \in \mathscr{M}, \Lambda_0 \in \mathscr{B}(\Omega_0)\}$,

$$\hat{X}_t(\omega) = \begin{cases} X_t(\omega), & \omega \in \Omega, \ t \geq 0 \\ x, & \omega = \omega_x \in \Omega_0, \ t \geq 0, \qquad (\hat{X}_\infty(\omega) = \Delta) \end{cases}$$

$$P_x(\Gamma) = \begin{cases} P_x(\Gamma \cap \Omega), & x \in S_\Delta, \ \Gamma \in \hat{\mathscr{M}} \\ I_{\{\omega_x\}}(\Gamma), & x \in \hat{S} - S, \ \Gamma \in \hat{\mathscr{M}}. \end{cases}$$ Q.E.D.

The last transform of $\mathbf{M}$ that we are concerned with is the operation of killing the sample paths upon leaving a Borel subset $\tilde{S} \subset S$. We set $S = \tilde{S} + B$ and

$$(4.1.8) \quad X_t^0(\omega) = \begin{cases} X_t(\omega) & 0 \leq t < \dot{\sigma}_B(\omega) \\ \Delta & t \geq \dot{\sigma}_B(\omega), \ \omega \in \Omega, \quad (X_\infty^0(\omega) = \Delta). \end{cases}$$

The quadruplet

$$(4.1.9) \quad \mathbf{M}_{\tilde{S}} = (\Omega, \mathscr{M}, X_t^0, \{P_x\}_{x \in \tilde{S}_\Delta})$$

is called the *part of the process* $\mathbf{M}$ *on the subset*. $\tilde{S}$.

Theorem 4.1.4. The part $\mathbf{M}_{\tilde{S}}$ of the Hunt process $\mathbf{M}$ on a relatively open subset $\tilde{S} \subset S$ is a standard process. When $\tilde{S}$ is a general Borel subset, $\mathbf{M}_{\tilde{S}}$ satisfies all the conditions of the standard process except that the normality condition (M. 5) may be violated and the Borel measurability condition (M. 2) holds in a weaker sense of $\mathscr{B}^*$-measurability.

Proof. The minimum completed admissible family $\{\mathscr{F}_t\}$ for the original process $\mathbf{M}$ is still admissible for $\mathbf{M}_{\tilde{S}}$ because $\{X_t^0 \in E\} = \{X_t \in E\} \cap \{\dot{\sigma}_B > t\} \in \mathscr{F}_t$, $E \in \mathscr{B}(\tilde{S})$. Let σ be an $\{\mathscr{F}_t\}$-stopping time. By the strong Markov property of $\mathbf{M}$ with respect to $\{\mathscr{F}_t\}$, we have for $\mu \in \mathscr{P}(S_\Delta)$ and $\Lambda \in \mathscr{F}_\infty$

(4.1.10) $P_\mu(\theta_\sigma^{-1}\Lambda|\mathscr{F}_\sigma) = P_{X\sigma}(\Lambda)$, P_μ-a.s.

Hence, for $\mu \in \mathscr{P}(\tilde{S}_\Delta)$ and $A \in \mathscr{B}(\tilde{S})$, $P_\mu(X^0_{\sigma+s} \in A|\mathscr{F}_\sigma) = P_\mu(X_{\sigma+s} \in A, \sigma + s < \sigma_B|\mathscr{F}_\sigma) = P_\mu(X_s\circ\theta_\sigma \in A,\ s < \sigma_B\circ\theta_\sigma, \sigma < \sigma_B|\mathscr{F}_\sigma) = I_{\{\sigma<\sigma_B\}}\cdot P_{X\sigma}(X_s \in A, s < \sigma_B) = P_{X^0_\sigma}(X^0_s \in A)$ P_x-a.e., yielding the strong Markov property of $\mathbf{M}_{\tilde{S}}$ with respect to $\{\mathscr{F}_t\}$. The life time of $\mathbf{M}_{\tilde{S}}$ is $\zeta^0 = \zeta \wedge \dot{\sigma}_B$. Let σ_n be $\{\mathscr{F}_t\}$-stopping times increasing to σ. Then $P_\mu(\sigma < \zeta^0) = P_\mu(\sigma < \zeta, \sigma < \dot{\sigma}_B) = P_\mu(\lim_{n\to\infty} X_{\sigma_n} = X_\sigma, \sigma < \zeta, \sigma < \dot{\sigma}_B) = P_\mu(\lim_{n\to\infty} X^0_{\sigma_n} = X^0_\sigma, \sigma < \zeta^0)$. Therefore, $\mathbf{M}_{\tilde{S}}$ is quasi-left continuous on $(0, \zeta^0)$ with respect to $\{\mathscr{F}_t\}$.

If $\tilde{S}$ is relatively open, then $B = S - \tilde{S}$ can be approximated by relatively open sets A_n such that $A_n \supset \bar{A}_{n+1}$, $\bigcap_{n=1}^{\infty} A_n = B$. Utilizing the quasi-lfet continuity of $\mathbf{M}$, we easily get

(4.1.11) $P_\mu(\lim_{n\to\infty} \sigma_{A_n} = \dot{\sigma}_B) = 1.$

Since the transition function of $\mathbf{M}_{\tilde{S}}$ is $P_x(X^0_t \in E) = P_x(X_t \in E,\ t < \dot{\sigma}_B)$ and σ_{A_n} is $\{\mathscr{F}^0_{t+}\}$-stopping time, we see from (4.1.11) that $\mathbf{M}_{\tilde{S}}$ satisfies the Borel measurability condition (M. 2). Q.E.D.

In introducing the notions of regular points, thin sets, fine topology and so on for the Hunt process $\mathbf{M}$ on $(S, \mathscr{B})$, a key role is played by the *Blumenthal* 0–1 *law* asserting that the field $\mathscr{F}_0$ is trivial:

(4.1.12) $P_x(\Lambda) = 0$ or 1 for $x \in S$ and $\Lambda \in \mathscr{F}_0$.

When $\Lambda \in \mathscr{F}^0_0$, this is immediate from the simple Markov property (Lemma 4.1.1) and the normality (M. 5). For $\Lambda \in \mathscr{F}_0$, take $\Lambda' \in \mathscr{F}^0_0$ such that $P_x(\Lambda\Delta\Lambda') = 0$. Then $P_x(\Lambda) = P_x(\Lambda')$, which yields (4.1.12).

A set $B \subset S_\Delta$ is called *nearly Borel measurable* if there exist for each $\mu \in \mathscr{P}(S_\Delta)$ Borel sets B_1, $B_2 \in \mathscr{B}_\Delta$ such that $B_1 \subset B \subset B_2$ and $P_\mu(X_t \in B_2 - B_1, \exists t \geqq 0) = 0$. The family of all nearly Borel measurable subsets of S is denoted by $\mathscr{B}^n$. Clearly, $\mathscr{B} \subset \mathscr{B}^n \subset \mathscr{B}^*$.

The hitting time σ_B of a nearly Borel set B is still $\{\mathscr{F}_t\}$-stopping time because, for each μ, the set $\{\sigma_B \leqq t\}$ differs from $\{\sigma_{B_1} \leqq t\} \in \mathscr{F}_t$ by a P_μ-negligible set for B_1 as above. In particular, $\{\sigma_B > 0\} \in \mathscr{F}_0$ and $P_x(\sigma_B > 0) = 0$ or 1 by (4.1.12). We say that x is a *regular point* of B in the former case and an *irregular point* in the latter case. The totality of the regular points of B is denoted by B^r. A set $A \subset S$ is called *finely open* if the set $S - A$ is thin at each point of A, i.e., for each $x \in A$ there exists a set $B = B(x) \in \mathscr{B}^n$ such that $B \supset S - A$ and $P_x(\sigma_B > 0) = 1$. Any relatively open set is finely open in view of the right continuity of the sample path. Hence, the *fine topology* defined by the finely open sets is finer than the original topology.

If a set A is contained in a set $B \in \mathscr{B}^n$ with $B^r = \phi$, then A is called *thin set*. A is called *semi-polor* if $A \subset \bigcup_n A_n$ for some thin sets A_n. One important fact in regard to this notion is that, for any $A \in \mathscr{B}^n$, the set $A - A^r$ is semi-polar ([14; II, (3,3)]).

If a set A is contained in a set $B \in \mathscr{B}^n$ with $P_x(\sigma_B < \infty) = 0$, $\forall x \in X$, then A is called *polar*. Obviously, any polar set is semi-polar but the converse is not true in general.

A Markovian transition function $\{p_t, t > 0\}$ on $(X, \mathscr{B}(X))$ is said to possess the *Feller property* if $p_t(C_\infty(X)) \subset C_\infty(X)$[1] and $\lim_{t \downarrow 0} p_t f(x) = f(x)$ $\forall x \in X$, $\forall f \in C_\infty(X)$. As is well known, $\{p_t, t > 0\}$ then admits a Hunt process on $(X, \mathscr{B}(X))$ ([14; I, (9, 4)]). Furthermore, if $\{p_t, t > 0\}$ satisfies the additional condition that for each compact set K and its neighbourhood G,

$$(4.1.13) \quad \lim_{t \downarrow 0} \sup_{x \in K} \frac{1}{t} p_t(x, X - G) = 0,$$

then the associated Hunt process $\mathbf{M}$ is a *diffusion* in the sense that

$$(4.1.14) \quad P_x(X_t \text{ is continuous in } t \in [0, \zeta)) = 1$$

for every $x \in X$ ([14; I, (9.10)]).

Example 4.1.1. (Spatially homogeneous Markov process).

Let a transition function $\{p_t, t > 0\}$ on $(R^n, \mathscr{B}(R^n))$ be given by (1.4.20) with a family $\{\nu_t, t > 0\}$ of probability measures satisfying (1.4.17) and (1.4.19). From the expression

$$(4.1.15) \quad p_t f(x) = \int_{R^n} f(x + y)\, \nu_t(dy), \; f \in C_\infty(R^n),$$

it is easy to see that $\{p_t, t > 0\}$ has the Feller property. Let $\mathbf{M} = (\Omega, \mathscr{M}, X_t, P_x)$ be the associated Hunt process on $(R^n, \mathscr{B}(R^n))$.

We may assume that $\zeta(\omega) = \infty$, $\forall \omega \in \Omega$, because $P_x(\zeta < t) = 1 - p_t 1(x) = 0$, $\forall x \in R^n$, $\forall t > 0$. $\mathbf{M}$ possesses the *spatial homogeneity:* $P_x(X_{t_1} \in E_1, \ldots, X_{t_k} \in E_k) = P_0(X_{t_1} + x \in E_1, \ldots, X_{t_k} + x \in E_k)$,

1) $C_\infty(X)$ is the family of continuous functions vanishing at infinity.

$t_1 < \ldots < t_k$, $E_1 \in \mathscr{B}(R^n), \ldots, E_k \in \mathscr{B}(R^n)$, that is, the law of P_x is obtained by the spatial shift from the single stochastic process $(\Omega, \mathscr{M}, X_t, P_0)$ starting at the origin 0. The latter is called a *Lévy process*, which possesses, together with the regularity (**M**.6) of sample paths, the property of stationary independent increments (K. Ito[47]).

The Hunt process on R^n with the transition function being given by the symmetric stable convolution semigroup ν_t of index α(see Example 1.5.2) is called the *symmetric stable process*. In particular, the (standard) *Brownian motion* on R^n is the Hunt process with ν_t being given by (1.4.25). This is a diffusion since the transition function satisfies the Lindeberg type condition (4.1.13).

§ 4.2 Smallness of Sets and Symmetry

Let X be a locally compact separable Hausdorff space and m be an everywhere dense positive Radon measure on X. We call a Markov process **M** on $(X, \mathscr{B}(X))$ *m-symmetric* if the transition function of **M** is m-symmetric in the sense of § 1.4. From now on throughout Chapter 4, we fix an m-symmetric Hunt process $\mathbf{M} = (\Omega, \mathscr{M}, X_t, P_x)$ on $(X, \mathscr{B}(X))$.

As indicated by the diagram at the end of this section, we study in this section the relationship among several notions of the smallness of sets related to the process **M** mainly based on the symmetry of **M** (or rather on the duality with respect to the measure m). The transition function of **M** is denoted by $\{p_t, t > 0\}$. The resolvent $\{R_\alpha, \alpha > 0\}$ of **M** is defined by

$$(4.2.1) \quad R_\alpha(x, E) = \int_0^\infty e^{-\alpha t} p_t(x, E)\, dt = E_x\Big(\int_0^\infty e^{-\alpha t} I_E(X_t) dt\Big).$$

The kernel R_α is m-symmetric as well. Both p_t and R_α are uniquely extended to a kernel on $(X, \mathscr{B}^*)$, $\mathscr{B}^*$ being the family of all universally measurable sets of X.

Lemma 4.2.1. The following conditions are equivalent each other for $E \in \mathscr{B}^*$:

(i) $m(E) = 0$,

(ii) $p_t(x, E) = 0$ m-a.e. $x \in X$ for each $t > 0$,

(iii) $R_\alpha(x, E) = 0$ m-a.e. $x \in X$ for each $\alpha > 0$.

Proof. We only show the implication (ii) $\Rightarrow$ (i). By the m-symmetry of p_t, $0 = \lim_{t \downarrow 0} \int_X p_t(x, E) m(dx) = \lim_{t \downarrow 0} \int_E p_t 1(x) m(dx) \geq m(E)$. Q.E.D.

A set $E \in \mathscr{B}^*$ is said to be *of potential zero* if $R_\alpha(x, E) = 0$, $^\forall x \in X$.

In particular, the above lemma means that any set in $\mathscr{B}^*$ of potential zero is m-negligible.

We have introduced three important notions of exceptional sets in the Hunt probabilistic potential theory—polar set, semi-polar set and set of potential zero. In this book, however, we shall use the term "exceptional set" exclusively in the following sense: a set $N \subset X$ is called *exceptional* if there exists a Borel set $\tilde{N} \supset N$ such that $P_m(\sigma_{\tilde{N}} < \infty) = 0$. Here $P_m(\Lambda) = \int_X P_x(\Lambda) m(dx)$, $\Lambda \in \mathscr{F}_\infty$.

Any exceptional set is m-negligible. In fact, for a Borel exceptional set N we have $p_t(x, N) = 0$ m-a.e. $x \in X$ and consequently $m(N) = 0$ by Lemma 4.2.1. A countable union of exceptional sets is also exceptional.

We say that a set $N \subset X$ is *properly exceptional* if N is Borel, $m(N) = 0$ and $X - N$ is $\mathbf{M}$-invariant. It is obvious that any properly exceptional set is exceptional. Conversely, it can be shown that any exceptional set is contained in a certain properly exceptional set. Before proving this, we explore some more implications of the m-symmetry of $\mathbf{M}$.

Lemma 4.2.2.

$$E_m(f_0(X_0)f_1(X_{t_1}) \cdot \cdot f_{n-1}(X_{t_{n-1}})\, f_n(X_{t_n}))$$
$$= E_m(f_n(X_0)\, f_{n-1}(X_{t_n - t_{n-1}}) \cdot \cdot \cdot f_1(x_{t_n - t_1})\, f_0(X_{t_n}))$$

where $0 < t_1 < \cdot \cdot \cdot < t_{n-1} < t_n$, $f_0, f_1, \cdot \cdot, f_n \in \mathscr{B}^+(X)$,[1)] and E_m denotes the integration on Ω with respect to the measure P_m.

Proof. Suppose the above equality holds for a given n. Then $E_m(f_0(X_0) \cdot \cdot f_n(X_{t_n}) f_{n+1}(X_{t_{n+1}})) = E_m(f_0(X_0) \cdot \cdot (f_n \cdot p_{t_{n+1}-t_n} f_{n+1})(X_{t_n})) = \int_X p_{t_{n+1}-t_n} f_{n+1}(x)\; E_x(f_n(X_0) f_{n-1}(X_{t_n - t_{n-1}}) \cdot \cdot f_0(X_{t_n})) m(dx)$.

The last expression equals, by virtue of the m-symmetry of $p_{t_{n+1}-t_n}$, $\int_X f_{n+1}(x)\; E_x(E_{X_{t_{n+1}-t_n}}(f_n(X_0) f_{n-1}(X_{t_n - t_{n-1}}) \cdot \cdot f_0(X_{t_n}))\, m(dx) = E_m\,(f_{n+1}(X_0)\, f_n(X_{t_{n+1}-t_n}) \cdot \cdot f_0(X_{t_{n+1}}))$, proving the desired equality for $n + 1$. Q.E.D.

For a Borel set $A \subset X$, we put

$$(4.2.2) \quad p_t^0(x, E) = P_x(X_t \in E,\ t < \sigma_A), \quad t > 0, \quad x \in X, \quad E \in \mathscr{B}(X).$$

This is a Markovian transition function on $(X, \mathscr{B}^*(X))$ in the sense of §1.4 and indeed the restriction of $\{p_t^0, t > 0\}$ to $(X - A, \mathscr{B}^*(X - A))$ is the transition function of the part $\mathbf{M}_{X-A}$ of the process $\mathbf{M}$ on $X - A$. By making use of Lemma 4.2.2, we can derive the m-symmetry of $\{p_t^0, t > 0\}$:

1) "$f \in \mathscr{B}^+(X)$" means that f is a non-negative Borel function on X. In Chapters 4 and 5, any function f on X is extended to X_Δ by setting $f(\Delta) = 0$.

Lemma 4.2.3. $\{p_t^0, t > 0\}$ is m-symmetric, i.e.,

$$(4.2.3)\quad \int_X f(x)p_t^0 g(x)m(dx) = \int_X g(x)p_t^0 f(x)m(dx), \quad f, g \in \mathscr{B}^+(X).$$

Proof. Assume first that A is open. By the previous lemma, we have $\int_X f(x)E_x(g(X_t); X_{kt/2n} \in X - A, 1 \leqq k \leqq 2^n, n = 1,2, \cdots)m(dx) = \int_X g(x)E_x(f(X_t); X_{kt/2n} \in X - A, 1 \leqq k \leqq 2^n, n = 1,2, \cdots)\ m(dx)$ for any $f, g \in C_0^+(X)$.[1] Hence,

$$(4.2.4)\quad \int_X f(x)\, E_x(g(X_t); t \leqq \sigma_A)m(dx) = \int_X g(x)E_x(f(X_t); t \leqq \sigma_A)m(dx)$$

in view of the regularity (M.6) of the sample paths and the closedness of $X - A$. Replacing t by $t + 1/n$ and letting $n \to \infty$, we arrive at (4.2.3).

Let A be an arbitrary Borel set. Consider a strictly positive function h such that $\int_X h(x)m(dx) = 1$. Note that any semi-polar set is of potential zero ([14; II (3.4)]) and consequently m-negligible. In particular, $h \cdot m$ does not charge the set $A - A^r$. Hence by the Hunt approximation theorem ([14; I, (11.3)]) there exists a decreasing sequence $\{G_n\}$ of open sets containing A such that $P_{h\cdot m}(\sigma_{G_n} \uparrow \sigma_A, n \uparrow \infty) = 1$. Therefore, $E_x(g(X_t); t < \sigma_{G_n})$ increases to $E_x(g(X_t); t < \sigma_A)$ m-a.e. $x \in X$ when $g \geqq 0$. Now the identity (4.2.3) for open sets leads us to the same one for the Borel A. Q.E.D.

Let us define the α-order *hitting distribution* H_α^A by

$$(4.2.5)\quad H_\alpha^A(x, E) = E_x(e^{-\alpha\sigma_A}\, I_E(X_{\sigma_A})),\ \alpha > 0,\ x \in X,\ E \in \mathscr{B}(X).$$

For a Borel set A, H_α^A is a Markovian kernel on $(X, \mathscr{B}^*(X))$. By making use of the strong Markov property, we easily get the so-called *Dynkin formula:*

$$(4.2.6)\quad R_\alpha f(x) = R_\alpha^0 f(x) + H_\alpha^A R_\alpha f(x),\ \ x \in X,\ f \in \mathscr{B}^+(X).$$

Here $\{R_\alpha^0, \alpha > 0\}$ is the resolvent of $\{p_t^0, t > 0\}$ defined by

$$(4.2.7)\quad R_\alpha^0(x, E) = \int_0^\infty e^{-\alpha t} p_t^0(x, E)dt = E_x\left(\int_0^{\sigma_A} e^{-\alpha t} I_E(X_t)dt\right).$$

1) $C_0^+(X)$ is the family of non-negative continuous functions on X with compact support.

As an immediate consequence of Lemma 4.2.3, we have

Corollary. For Borel A and $\alpha > 0$, the identity

$$(4.2.8)\quad \int_X f(x)H_\alpha^A R_\alpha g(x)m(dx) = \int_X g(x)H_\alpha^A R_\alpha f(x)m(dx)$$

holds for any non-negative universally measurable f and g.

Lemma 4.2.4. Let A be nearly Borel and finely open. If A is m-negligible, then A is exceptional.

Proof. Let K be a compact subset of A. By (4.2.8) we have for any $f \in C_0(X)$[1])

$$\int_X f(x)H_\alpha^K R_\alpha I_A(x)\mathrm{m}(dx) = \int_A H_\alpha^K R_\alpha f(x)m(dx) = 0.$$

Therefore, $H_\alpha^K R_\beta I_A(x) = 0$, $\beta > \alpha$, for m-a.e. $x \in X$. Since the hitting distribution $H_\alpha^K(x, \cdot)$ is supported by K and $\lim_{\beta\to\infty} \beta R_\beta I_A(y) = 1$, $^\forall y \in K \subset A$, we get $H_\alpha^K 1(x) = E_x(e^{-\alpha\sigma_K}) = 0$ m-a.e., which means that K is exceptional.

Take now a strictly positive function h such that $\int_X h(x)m(dx) = 1$. By the Hunt approximation theorem [14; I, (10.19)], there exists an increasing sequence $\{K_n\}$ of compact subsets of A such that $E_{h\cdot m}(e^{-\alpha\sigma_A}) = \lim_{n\to\infty} E_{h\cdot m}(e^{-\alpha\sigma_{K_n}}) = 0$. Therefore, A is exceptional. Q.E.D.

We use the term *"quasi-everywhere"* or "*q.e.*" to mean "except on an exceptional set". A function u defined q.e. on X is called *finely continuous q.e.* if there exists a nearly Borel exceptional set N such that $X - N$ is finely open and u is nearly Borel and finely continuous on $X - N$.

Lemma 4.2.5. If u is finely continuous q.e. and $u \geqq 0$ m-a.e., then $u \geqq 0$ q.e.

Proof. Let N be the set appearing in the above definition of q.e. fine continuity. Then the set $A = \{x \in X - N\colon u(x) < 0\}$ is nearly Borel, finely open and m-negligible. Hence, A is exceptional by Lemma 4.2.4. Q.E.D.

1) Cf. §1.1.

A universally measurable function u on X is said to be *α-excessive* if $u(x) \geqq 0$, $e^{-\alpha t}p_t u(x) \uparrow u(x)$, $t \downarrow 0$, $x \in X$ $(\alpha > 0)$. A 0-excessive function is simply called *excessive*. It is known that u is then nearly Borel measurable and finely continuous ([14; II (2.11) and (4.2)]). A typical example of an α-excessive function is $u = R_\alpha f$ for non-negative universally measurable f. If u is α-excessive, then so is its α-harmonic average $H_\alpha^A u$ relative to a nearly Borel A. We set

$$(4.2.9) \quad \begin{cases} p_A^\alpha(x) = H_\alpha^A 1(x) \; (= E_x(e^{-\alpha\sigma_A})) \\ p_A(x) = H_{0+}^A 1(x) \; (= P_x(\sigma_A < \infty)). \end{cases}$$

p_A^α (resp. p_A) is an α-excessive (resp. excessive) function. Keeping these in mind we proceed to the proof of the following theorem.

Theorem 4.2.1 A set is exceptional if and only if it is contained in a properly exceptional set.

Proof. If N is exceptional, then there is a Borel set $B_0 \supset N$ such that $P_{B_0}(x) = 0$ m-a.e. By virtue of Lemma 4.2.5, p_{B_0} vanishes q.e., that is, except on some Borel exceptional set B_1. Apply the same argument to the function $p_{B_0 \cup B_1}$. In this way we get a sequence $\{B_k\}_{k=0}^{\infty}$ of Borel exceptional sets such that $p_{(\bigcup_{k=1}^{n} B_k)}(x) = 0$, $\forall x \in X - B_{n+1}$. We thus get $p_B(x) = 0$, $\forall x \in X - B$, for the Borel exceptional set $B = \bigcup_{k=0}^{\infty} B_k$.

On the other hand, by making use of the Hunt approximation theorem ([14; I, (11.3)]) again, we can readily observe that $p_B(x) = p_x(X_t$ or $X_{t-} \in B, \exists t > 0)$, $x \in X - B$. Hence, $X - B$ is M-invariant and B is properly exceptional. Q.E.D.

We can now simplify the description of q.e. fine continuity as follows:

Lemma 4.2.6. A function u is finely continuous q.e. if and only if there exists a properly exceptional set B such that u is Borel measurable and finely continuous on $X - B$.

Proof. We prove this when u is bounded. Let N be the set appearing in the definition of q.e. fine continuity for u. Then $\lim_{n\to\infty} nR_n u(x) = u(x)$, $\forall x \in X - N$. We extend u to X by setting $u(x) = 0$, $x \in N$. Choose Borel u_1 and u_2 such that $u_1 \leqq u \leqq u_2$ on X and $u_1 = u_2$ m-a.e. Then by the m-symmetry of R_n, we see that $R_n u_1 = R_n u_2$ m-a.e. and consequently q.e. on account of Lemma 4.2.5. Let B_n be the corresponding exceptional set. Then there exists by virtue of Thoerem 4.2.1 a properly exceptional set B containing N and all B_n, $n = 1,2, \ldots$. On $X - B$ we have $u(x) = \lim_{n\to\infty} nR_n u(x) = \lim_{n\to\infty} nR_n u_1(x)$, which yields the Borel measurability of u. Q.E.D.

We state here a present version of [14; VI, (3.2)]. The proof is omitted because it is quite the same.

Lemma 4.2.7. Let $\{u_n\}$ be a decreasing sequence of α-excessive functions with limit u and suppose that $u = 0$ m-a.e. Then $u = 0$ q.e.

Any polar set A is exceptional. To see this, let B a nearly Borel polar set containing A. Taking a strictly positive function h such that $\int_X h(x)m(dx) = 1$, choose Borel sets B_1, B_2 such that $B_1 \subset B \subset B_2$ and $\sigma_{B_2 - B_1} = \infty$ $P_{h\cdot m}$-a.s. Then clearly $\sigma_{B_2} = \infty$ $P_{h\cdot m}$-a.s. and hence A is exceptional.

We claim that the converse statement holds if and only if the process **M** satisfies an absolute continuity condition.

Theorem 4.2.2. The following two conditions are equivalent:
(i) A set is polar if and only if it is exceptional,
(ii) $R_\alpha(x, \cdot)$ is absolutely continuous with respect to m for each $\alpha > 0$ and $x \in X$.

Proof. (i) $\Rightarrow$ (ii): Suppose E is Borel and $m(E) = 0$, then, in view of Lemma 4.2.1, $R_\alpha(x, E) = 0$ for m-a.e. $x \in X$ and consequently q.e. $x \in X$ by virtue of Lemma 4.2.5. Let N be the corresponding Borel exceptional set. N is then polar on account of (i) and $R_\alpha(x, E) = \lim_{t\downarrow 0} E_x(R_\alpha(X_t, E); X_t \notin N) = 0$, $\forall x \in X$.
(ii) $\Rightarrow$ (i): If B is a Borel exceptional set, then $u(x) = P_x(\sigma_B < \infty) = 0$ m-a.e. u being excessive, we have by condition (ii) that $u(x) = \lim_{\alpha\to\infty} \alpha R_\alpha u(x) = 0$ $\forall x \in X$. Q.E.D.

The last theorem in this section concerns the semi-polar set. We need an amplified version of (4.2.8): for any Borel sets A and B

$$(4.2.10)\quad (g,\ H_\alpha^A H_\alpha^B R_\alpha h) = (H_\alpha^B H_\alpha^A R_\alpha g,\ h), \quad g,\ h \in \mathscr{B}^+.$$

In fact, this follows from (4.2.8) and the symmetry of R_α, for $(g, H_\alpha^A H_\alpha^B R_\alpha h) = \lim_{\beta\to\infty} \beta(g, H_\alpha^A R_\beta H_\alpha^B R_\alpha h) = \lim_{\beta\to\infty} \beta(g, H_\alpha^A R_\alpha(I - (\beta-\alpha)R_\beta)H_\alpha^B R_\alpha h) = \lim_{\beta\to\infty} \beta(H_\alpha^B R_\alpha(I - (\beta - \alpha)R_\beta)H_\alpha^A R_\alpha g,\ h) = \lim_{\beta\to\infty} \beta(H_\alpha^B R_\beta H_\alpha^A R_\alpha g,\ h) = (H_\alpha^B H_\alpha^A R_\alpha g,\ h)$. As an easy consequence of (4.2.10), we get

(4.2.11) $H_\alpha^G p_K^\alpha(x) = p_K^\alpha(x)$ m-a.e.

for any compact set K and an open set $G \supset K$.

Theorem 4.2.3 Any semi-polar set is exceptional.

Proof. Let K be an arbitrary compact thin set. It suffices to prove that K is exceptional, because then the general statement of the theorem follows in the same way as in the proof of Lemma 4.2.4.

First note that

(4.2.12) $P_x(\lim_{t \uparrow \sigma_K} p_K^\alpha(X_t) = 1, \sigma_K < \infty) = P_x(\sigma_K < \infty)$ m-a.e. $x \in X$.

To see this, let $\{G_n\}$ be open sets such that $G_n \supset \bar{G}_{n+1}$ and $\bigcap_n G_n = K$. From (4.2.11), we see $\sigma_{G_n} + \sigma_K \circ \theta \sigma_{G_n} = \sigma_K$. In view of the thinness of K and identity (4.1.11), we then have $P_x(\sigma_{G_n} < \sigma_K, \forall n, \lim_{n \to \infty} \sigma_{G_n} = \sigma_K, \sigma_K < \infty) = P_x(\sigma_K < \infty)$ m-a.e. $x \in X$, which, combined with (4.2.11), means (4.2.12).[1)]

Now set $B_n = \{x \in X : p_K^\alpha(x) \geq 1 - 1/n\}$. Since each B_n is finely closed and $\bigcap_{n=1}^\infty B_n$ is empty, (4.2.12) implies that $\sigma_{B_n} < \sigma_K$ P_x-a.s. on $\{\sigma_K < \infty\}$ for m-a.e. $x \in X$. Hence, $H_\alpha^{B_n} H_\alpha^K f = H_\alpha^K f$ m-a.e. $\forall f \in \mathscr{B}^+(X)$, and (4.2.10) together with (4.2.8) leads us to

(4.2.13) $(H_\alpha^K g, h) = (H_\alpha^K H_\alpha^{B_n} g, h)$.

Take a strictly positive L^1-function h; then (4.2.13) means that the measure $\pi(E) = (H_\alpha^K I_E, h)$ is concentrated on each B_n. Since $\bigcap_{n=1}^\infty B_n$ is empty, $\pi = 0$ and consequently $p_K^\alpha(x) = H_\alpha^K 1(x) = 0$ m-a.e., proving that K is exceptional. Q.E.D.

The following diagram indicates the relations among small sets in the m-symmetric Hunt process **M**:

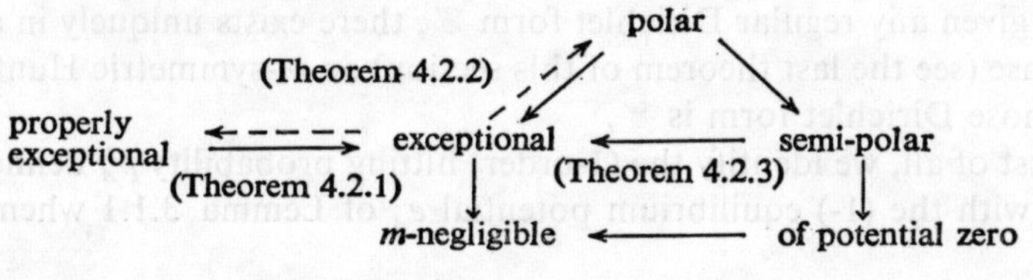

Example 4.2.1. The spatially homogeneous Markov process on R^n (see

1) Cf. [14; II, (3.12)].

Example 4.1.1) is symmetric with respect to the Lebesgue measure if and only if $\{\nu_t\}$ satisfies (1.4.18). If, moreover, ν_t is absolutely continuous with respect to the Lebesgue measure (for instance if **M** is the Brownian motion or the symmetric stable process), then the notion of exceptional set is merely a synonym for polar set by Theorem 4.2.2. In this case, any Borel polar set is also properly exceptional.

However, ν_t is not absolutely continuous in general. In the one dimensional case, for example, it is known that ν_t is purely discontinuous if and only if $S = 0$ and J is purely discontinuous and bounded in the Lévy-Khinchin formula (1.4.21).[1] Thus, the role of the present notion of the exceptional set can not be covered by the traditional notion of the polar set in general (see Theorem 4.3.4).

Theorem 4.2.3 is violated by some non-symmetric Markov process. Take the deterministic process **M** of uniform motion to the right: **M** has the transition function

(4.2.14) $$p_t(x, A) = \delta_{\{x+t\}}(A), \ t > 0, \ x \in R^1, \ A \in \mathscr{B}(R^1).$$

Then every singleton is thin but not exceptional.

§4.3 Identification with Potential Theoretic Notions

The transition function $\{p_t, t > 0\}$ of the m-symmetric Hunt process **M** on X clearly satisfies the continuity condition (1.4.15). Hence, $\{p_t, t > 0\}$ uniquely determines a strongly continuous Markovian semigroup $\{T_t, t > 0\}$ on $L^2(X; m)$ and a Dirichlet form $\mathscr{E}$ on $L^2(X; m)$ as in Lemma 1.4.3 and the paragraph immediately following it respectively. We call them respectively the *semigroup* and the *Dirichlet form* on $L^2(X; m)$ *of the process* **M**.

In this section, several probabilistic notions which have appeared in the preceding sections relevent to **M** are identified with those analytical notions in Chapter 3 relevent to $\mathscr{E}$. In doing so, we need the following assumption (4.3.1) for the process **M**:

(4.3.1) *the Dirichlet form $\mathscr{E}$ on $L^2(X; m)$ of* **M** *is regular.*

This condition on **M** is not very restrictive because we show in Chapter 6 that, given any regular Dirichlet form $\mathscr{E}$, there exists uniquely in a certain sense (see the last theorem of this section) an m-symmetric Hunt process whose Dirichlet form is $\mathscr{E}$.

First of all, we identify the (1-order) hitting probability p_A^1 defined by (4.2.9) with the (1-) equilibrium potential e_A of Lemma 3.1.1 when A is open.

---
1) K.Ito [48; §0.5].

Lemma 4.3.1. If A is an open set with finite capacity, then p_A^1 is a version of e_A.

Proof. As T_t is the extension to $L^2(X; m)$ of p_t on $\mathscr{B}_b \cap L^2(X; m)$[1], $p_t f$ is a version of $T_t f$ when f is a universally measurable non-negative function in $L^2(X; m)$. In particular, if f is α-excessive with respect to $\{p_t, t > 0\}$, then it is also α-excessive with respect to $\{T_t, t > 0\}$ in the sense of § 3.2.

Since p_A^1 is 1-excessive with respect to $\{p_t, t > 0\}$ and $p_A^1(x) = 1$, $\forall x \in A$, it suffices to show, in view of Lemma 3.1.1 and Lemma 3.3.2 that

(4.3.2) $\quad p_A^1 \leqq e_A \quad m\text{-a.e.}$

Take a Borel modification $\tilde{e}_A$ of e_A such that $\tilde{e}_A(x) = 1$, $\forall x \in A$, and let $Y_t(\omega) = e^{-t}\tilde{e}_A(X_t(\omega))$, $t \geqq 0$, $\omega \in \Omega$. Then $(Y_t, \mathscr{F}_t^0, P_{h\cdot m})_{t \geqq 0}$ is a supermartingale[2] for any non-negative Borel function h such that $\int_X h(x)m(dx) = 1$. Indeed, the Markov property of $\mathbf{M}$ implies, for $0 \leqq s < t$, $E_{h\cdot m}(e^{-t}\tilde{e}_A(X_t) | \mathscr{F}_s^0) = e^{-s}e^{-(t-s)}p_{t-s}\tilde{e}_A(X_s)$, $P_{h\cdot m}$-a.s., which in turn is not greater than $e^{-s}\tilde{e}_A(X_s)$, $P_{h\cdot m}$-a.s., because the set $\{x \in X\colon e^{-(t-s)} p_{t-s}\tilde{e}_A(x) > \tilde{e}_A(x)\}$ is m-negligible.

Let D be a finite subset of $(0, \infty)$ with min $D = a$ and max $D = b$ and put $\sigma(D; A) = \min \{t \in D;\ X_t \in A\}$. If the set in the braces is empty, we set $\sigma(D; A) = b$. The optional sampling theorem ([65; V, T28]) applied to $\{Y_t, \mathscr{F}_t^0, P_{h\cdot m}\}_{t \in D}$ yields

$$E_{h\cdot m}(e^{-\sigma(D;A)};\ \sigma(D; A) < b) \leqq E_{h\cdot m}(Y_{\sigma(D;A)}) \leqq E_{h\cdot m}(Y_a) \leqq (h, \tilde{e}_A).$$

Letting D increase to a countable dense subset of $(0, b)$ and then b tend to infinity, we arrive at $(h, p_A^1) \leqq (h, e_A)$ which means (4.3.2). Q.E.D.

The next theorem justifies the usage of the term "*q.e.*" in either the sense of § 3.1 or § 4.2.

Theorem 4.3.1. A set $N \subset X$ is exceptional if and only if $\mathrm{Cap}(N) = 0$.

Proof. Suppose $\mathrm{Cap}(N) = 0$. Then there exists a decreasing sequence $\{A_n\}$ of open sets such that $A_n \supset N$ and $\lim_{n\to\infty} \mathrm{Cap}(A_n) = 0$. Since $\mathrm{Cap}(A_n) = \mathscr{E}_1(e_{A_n}, e_{A_n}) \geqq (e_{A_n}, e_{A_n})$, we have from Lemma 4.3.1 that

1) $\mathscr{B}_b$ denotes the set of bounded Borel functions.
2) Cf. P.A.Meyer [65].

(4.3.3) $\lim_{n\to\infty} p^1_{A_n}(x) = 0$ m-a.e.

and, consequently, $p^1_B(x) = 0$ m-a.e. for $B = \bigcap_{n=1}^{\infty} A_n (\supset N)$. Hence, N is exceptional.

Conversely, assume that N is a compact exceptional set. Let $\{G_n\}$ be a sequence of decreasing, relatively compact open sets such that $G_n \supset \bar{G}_{n+1}$ and $\bigcap_{n=1}^{\infty} G_n = N$. Then $\lim_{n\to\infty} p^1_{G_n}(x) = p^1_N(x) = 0$ m-a.e. because of (4.1.11) and m-negligibility of N. Since $\mathrm{Cap}(G_n) < \infty$, Lemma 4.3.1 and same reasoning as in the proof of Lemma 3.1.2 (iii) imply that $\mathrm{Cap}(N) \leqq \lim_{n\to\infty} \mathrm{Cap}(G_n) = \lim_{n\to\infty} \mathscr{E}_1(p^1_{G_n}, p^1_{G_n}) = 0$. Now it suffices to use (3.1.6) to complete the proof. Q.E.D.

We now identify the notions of the continuity of functions introduced in § 3.1 and § 4.2.

Theorem 4.3.2. (i) If u is quasi-continuous, then u is finely continuous q.e. More precisely, there exists a properly exceptional set N such that u is Borel measurable on $X - N$ and for any $x \in X - N$,

(4.3.4) $P_x(u(X_t)$ is right continuous and $\lim_{t'\uparrow t} u(X_{t'}) = u(X_{t-}), \forall t \geqq 0) = 1.$

(ii) Conversely, if u is finely continuous q.e. and $u \in \mathscr{F}$, then u is quasi-continuous.

Proof. (i) There exists a decreasing sequence $\{A_n\}$ of open sets such that $\lim_{n\to\infty} \mathrm{Cap}(A_n) = 0$ and $u|_{X-A_n}$ is continuous for each n. Then we have the relation (4.3.3) which, combined with Lemma 4.2.7, can be strengthened as

(4.3.5) $\lim_{n\to\infty} p^1_{A_n}(x) = 0$ q.e.

Denote by N_0 the exceptional set in (4.3.5).

Due to Theorem 4.2.1, we can find a properly exceptional set N containing $N_0 \cup (\bigcap_{n=1}^{\infty} A_n)$. The property (4.3.4) clearly holds for $x \in X - N$ because $P_x(\lim_{n\to\infty} \sigma_{A_n} = \infty) = 1$, $x \in X - N$. Therefore, u is finely continuous on $X - N$ ([14; II, (4.8)]).

(ii) Let $u \in \mathscr{F}$ be finely continuous q.e. By Theorem 3.1.3, u admits a quasi-continuous version $\tilde{u}$. $\tilde{u}$ is then finely continuous q.e. by (i). Since $u = \tilde{u}$ m-a.e., we conclude from Lemma 4.2.5 that $u = \tilde{u}$ q.e. As u differs from a quasi-continuous function $\tilde{u}$ only by a set of zero capacity, u is quasi-continuous. Q.E.D.

The next theorem asserts that the process **M** is *properly associated with* its Dirichlet form $\mathscr{E}$ in a sense.

Thoerem 4.3.3. For any non-negative universally measurable function $u \in L^2(X; m)$,
(i) $p_t u$ is a quasi-continuous version of $T_t u$, $t > 0$,
(ii) $R_\alpha u$ is a quasi-continuous version of $G_\alpha u$, $\alpha > 0$.

We prove this theorem utilizing the following variant of the monotone lemma.

Lemma 4.3.2. Let H be a family of non-negative functions on X such that
(H.1) $u_1, u_2 \in H$, $c_1 u_1 + c_2 u_2 \geqq 0$ for a constants c_1, $c_2 \Rightarrow c_1 u_1 + c_2 u_2 \in H$,
(H.2) $u_n \in H$, $u_n \uparrow u \in L^2(X; m) \Rightarrow u \in H$
(H.3) $\forall A$ open, $\exists u_n \in H$, $u_n \uparrow I_A$.
Then H contains all non-negative Borel functions in L^2 $(X; m)$.

Proof. Fix a relatively open set $E \subset X$ and put $\mathscr{D} = \{A \subset E$: A is Borel and $I_A \in H\}$. Then $\mathscr{D}$ is a Dynkin class containing all open subsets of E. Hence, $\mathscr{D} = \mathscr{B}(E)$ by [14; 0, (2.2)]. The rest of the proof is clear. Q.E.D.

Proof of Theorem 4.3.3. We already noticed that $p_t u$ is a version of $T_t u$. Since

$$(R_\alpha u, v) = \int_0^\infty e^{-\alpha t}(p_t u, v)dt = \int_0^\infty e^{-\alpha t}(T_t u, v)dt = (G_\alpha u, v), \ \forall v \in C_0(X),$$

$R_\alpha u$ is a version of $G_\alpha u \in \mathscr{F}$. Furthermore, $R_\alpha u$ is α-excessive and, consequently, finely continuous on X. Hence, we get (ii) by virtue of Theorem 4.3.2.

We next prove (i) for non-negative $u \in C_0(X)$. Then $R_\alpha p_t u$ is a quasi-continuous version of $G_\alpha T_t u$ and $\lim_{\alpha\to\infty} \alpha R_\alpha p_t u(x) = \lim_{\alpha\to\infty} p_t(\alpha R_\alpha)u(x) = p_t u(x)$, $\forall x \in X$. On the other hand, $\alpha G_\alpha T_t u$ is $\mathscr{E}_1$-convergent to $T_t u \in \mathscr{F}$

on account of Lemma 1.3.3. Therefore, $p_t u$ is a quasi-continuous version of $T_t u$ by virtue of Theorem 3.1.4 (ii).

Let us put $H = \{u \in L^2(X; m)$: u is non-negative Borel and satisfies (i)$\}$. We have just proved $H \supset C_0^+(X)$ and, hence, H satisfies $(H.3)$ of Lemma 4.3.2. $(H.1)$ is trivially satisfied. To check $(H.2)$, let $u_n \in H$ inecrease to $u \in L^2(X; m)$. Then $T_t u_n$ is $\mathscr{E}_1$-convergent to $T_t u$ by (i) of Lemma 1.3.3. Since $p_t u_n$ is a quasi-continuous version of $T_t u_n$ and convergent to $p_t u$ pointwise, we get $u \in H$ by Theorem 3.1.4 (ii) again.

Lemma 4.3.2 implies that (i) holds for u Borel. The extension to the universally measurable u is clear. Q.E.D.

As a result of the regularity of the transition function embodied in Theorem 4.3.3, we can prove the following:

Theorem 4.3.4. The equivalent two conditions of Theorem 4.2.2 are also equivalent to the following one: $p_t(x, \cdot)$ is absolutely continuous with respect to m for each $t > 0$ and $x \in X$.

Proof. Since this condition implies the second one of Theorem 4.2.2, it is sufficient to derive this property of p_t from the first condition (i) of Theorem 4.2.2.

Assume that any exceptional set is polar. If A is an m-negligible Borel set, then $p_t(x, A) = 0$ m-a.e. by Lemma 4.2.1. Since $p_t(x,A)$ is quasi-continuous as a function of $x \in X$ by Theorem 4.3.3, it vanishes q.e. by Theorem 3.1.2. Hence, there exists a Borel polar set N such that $p_t(x, A) = 0$ $\forall x \in X-N$. Therefore, $p_{2t}(x,A) = E_x(p_t(X_t, A); X_t \notin N) = 0$, $\forall x \in X$, and we obtain Theorem 4.3.4. Q.E.D.

Finally, we amplify the first lemma of this section as follows. Let e_B be the (1-) equilibrium potential of a Borel set B introduced in § 3.3. Together with the α-order hitting probability $p_B^\alpha(x) = E_x(e^{-\alpha\sigma_B})$, we also consider the function $\dot{p}_B^\alpha(x) = E_x(e^{-\alpha\dot{\sigma}_B})$ for the hitting time $\dot{\sigma}_B$ defined by (4.1.5).

Theorem 4.3.5. Let B be a Borel set of finite capacity. Then both p_B^1 and $\dot{p}_B^1$ are quasi-continuous versions of e_B.

Proof. When B is an open set of finite capacity, $p_B^1 = \dot{p}_B^1$ and p_B^1 is a version of e_B by Lemma 4.3.1. p_B^1 being 1-excessive, it is a finely continuous function belonging to $\mathscr{F}$ and, consequently, quasi-continuous.

If B is a Borel set of finite capacity, then by the proof of Theorem

3.3.1 there is a decreasing sequence $\{A_n\}$ of open sets of finite capacity and including B such that e_{A_n} is $\mathscr{E}_1$-convergent to e_B. Set $q(x) = \lim_{n\to\infty} p^1_{A_n}(x)$, $x \in X$; then

(4.3.6) $$p^1_B(x) \leq \dot{p}^1_B(x) \leq q(x), \quad x \in X.$$

q is a quasi-continuous version of e_B by virtue of Theorem 3.1.4 (ii).

We first conclude from inequality (4.3.6) and Lemma 3.3.2 that $p^1_B \in \mathscr{F}$. Theorem 4.3.2 then implies that p^1_B is quasi-continuous. Furthermore, $p^1_B = 1$ q.e. on B because of Theorem 4.2.3. Therefore p^1_B is a quasi-continuous version of e_B by Theorem 3.3.3(iii). The same property of $\dot{p}^1_B$ is clear from (4.3.6). Q.E.D.

So far we have fixed an m-symmetric Hunt process $\mathbf{M}$ on X whose Dirichlet form $\mathscr{E}$ on $L^2(X; m)$ is regular. Theorem 4.3.3. enables us to formulate a statement concerning the uniqueness of such a process.

Let us say that two m-symmetric Hunt processes are *equivalent* if they possess a common properly exceptional set outside which their transition functions coincide.

Theorem 4.3.6. Let $\mathbf{M}_1$ and $\mathbf{M}_2$ be two m-symmetric Hunt processes on X possessing a common regular Dirichlet space on $L^2(X; m)$. Then $\mathbf{M}_1$ and $\mathbf{M}_2$ are equivalent.

Proof. Denote by $p_t^{(i)}$ the transition function of $\mathbf{M}_i$, $i = 1, 2$. By virtue of Theorem 4.3.3 and Lemma 3.1.4, there exists a set B_0 of capacity zero such that

(4.3.7) $$p_t^{(1)}u(x) = p_t^{(2)}u(x), \quad \forall x \in X - B_0$$

for every $t \in Q_+$ and $u \in C_1(X)$, $C_1(X)$ being a countable uniformly dense subset of $C_0(X)$. (4.3.7) then holds for every $t > 0$ by right continuity. Applying Theorem 4.2.1 and Theorem 4.3.1 to $\mathbf{M}_1$ and $\mathbf{M}_2$ alternatively, we can find an increasing sequence $\{B_n\}$ of Borel sets containing B_0 such that B_{2n-1} (resp. B_{2n}) is properly exceptional relative to $\mathbf{M}_1$ (resp. $\mathbf{M}_2$). Then $B = \bigcup_{n=1}^{\infty} B_n$ is a common properly exceptional set outside which $p_t^{(1)} = p_t^{(2)}$, $\forall t > 0$. Q.E.D.

Example 4.3.1. Let $\mathbf{M}$ be the spatially homogeneous symmetric Markov process on R^n (see Example 4.2.1). Then the Dirichlet form of $\mathbf{M}$ is given by (1.4.24) and $\mathbf{M}$ satisfies the condition (4.3.1) according to Example 1.4.1.

In particular, the Dirichlet form $\mathscr{E}$ of the transient symmetric stable process **M** of index α is given by (1.4.24) with $\psi(x) = c|x|^\alpha$, $0 < \alpha \leqq 2$, $\alpha < n$, $c > 0$. On account of Theorem 3.1.5 and Example 3.3.1, a set N is of capacity zero with respect to $\mathscr{E}_1$ if and only if the set N is of capacity zero in the sense of classical potential theory (namely, relevent to the Riesz kernel $R^{(\alpha)}$ of (1.5.15)). Therefore, Theorem 4.3.1 and Example 4.2.1 imply that a set N is of capacity zero in the sense of the classical potential theory if and only if N is polar with respect to the stable process **M**.

Thus, the Brownian polar set is just the set of Newtonian capacity zero ($n \geqq 3$). Since the Brownian paths X_t are continuous (see Example 4.1.1), it follows from Theorem 3.1.5, Example 3.1.1 and Theorem 4.3.2 that any BLD function u of potential type on R^n has the following property:

(4.3.8) $P_x(u(X_t)$ is continuous in $t \in [0,\infty)) = 1$, $\forall x \in R^n - N$,

N being a certain Borel polar set depending on u.

Theorem 4.3.4 is violated by the uniform motion to the right: it has the transition function (4.2.14) which is not absolutely continuous with respect to Lebesgue measure. But the resolvent $R_\alpha(x, dy)$ has the density

$$R_\alpha(x, y) = \begin{cases} e^{\alpha(x-y)} & x < y \\ 0 & x \geqq y. \end{cases}$$

§4.4 Hitting Times and Orthogonal Projections

We still consider the m-symmetric Hunt process **M** on X satisfying condition (4.3.1). Just as in §3.3 we set for a fixed Borel set $B \subset X$

$$(4.4.1) \quad \mathscr{F}_{X-B} = \{u \in \mathscr{F}: \tilde{u} = 0 \quad \text{q.e.} \quad \text{on } B\},$$

$\tilde{u}$ being a quasi-continuous version of u. $\mathscr{F}_{X-B}$ is a closed subspace of the Hilbert space $(\mathscr{F}, \mathscr{E}_\alpha)$ for each $\alpha > 0$. Denote by $\mathscr{H}_\alpha^B$ its orthogonal complement:

$$(4.4.2) \quad \mathscr{F} = \mathscr{F}_{X-B} \oplus \mathscr{H}_\alpha^B.$$

The orthogonal projection on the space $\mathscr{H}_\alpha^B$ is denoted by $\mathbf{P}_{\mathscr{H}_\alpha^B}$.

The α-order hitting distribution H_α^B has been defined by (4.2.5). In dealing with a function v defined and non-negative q.e. on **X**, we always assume that v is extended to be a non-negative universally measurable function on X. Then the operation $H_\alpha^B v(x)$ makes sense for every $x \in X$ and $H_\alpha^B v(x) = E_x(e^{-\alpha\sigma_B} v(X_{\sigma_B}))$.

The key theorem of this section is the following:

Theorem 4.4.1. For each non-negative $u \in \mathscr{F}$, $H_\alpha^B \tilde{u}$ is a quasi-continuous version of $\mathbf{P}_{\mathscr{H}_\alpha^B} u$.

First, we shall prove a lemma.

Lemma 4.4.1. Let $u \in \mathscr{F}$ be an α-excessive function with respect to $\{p_t, t > 0\}$. Denote by u_B the α-reduced function of u on B defined at the end of §3.3. Then $H_\alpha^B u$ is a quasi-continuous version of u_B.

Proof. As we saw in the preceding section, u is quasi-continuous and α-excessive with respect to $\{T_t, t > 0\}$. It suffices to show

(4.4.3) $H_\alpha^B u \leqq u_B$ m-a.e.

(4.4.4) $H_\alpha^B u(x) = u(x)$ q.e. on B.

In fact, $H_\alpha^B u$ being α-excessive and finely continuous, we see from (4.4.3) and Lemma 3.3.2 that $H_\alpha^B u \in \mathscr{F}$. Then Theorem 4.3.2 (ii), (4.4.4), and (3.3.11) lead us to $\widetilde{H_\alpha^B u} = H_\alpha^B u = u = \tilde{u}_B$ q.e. on B. Thus, we can conclude $H_\alpha^B u = u_B$ by Theorem 3.3.3. Since (4.4.4) follows immediately from Theorem 4.2.3, it only remains to verify the inequality (4.4.3).

To prove (4.4.3) we take a non-negative Borel quasi-continuous version $\tilde{u}_B$ of u_B. We claim that there exists a properly exceptional set N such that

(4.4.5) $e^{-\alpha t} p_t \tilde{u}_B(x) \leqq \tilde{u}_B(x), \quad \forall t > 0, \quad \forall x \in X - N,$

(4.4.6) $\tilde{u}_B(x) = u_B(x), \ \forall x \in B - N,$

(4.4.7) $P_x(\tilde{u}_B(X_t)$ is right continuous in $t \geqq 0) = 1, \ x \in X - N.$

Since u_B is α-excessive with respect to $\{T_t, t > 0\}$ and $p_t \tilde{u}_B$ is quasi-continuous by Theorem 4.3.3, we see that (4.4.5) holds q.e. $x \in X$ for each $t > 0$. By the definition of $\tilde{u}_B$, (4.4.6) is true for q.e. $x \in B$. Furthermore, Theorem 4.3.2 (i) asserts the validity of (4.4.7) for q.e. $x \in X$. According to Theorem 4.2.1, we can choose a properly exceptional set N such that the inequality (4.4.5) for every rational $t > 0$, the equality (4.4.6) and the relation (4.4.7) hold for all $x \in X - N$. An application of Fatou's lemma then yields (4.4.5) for every $t > 0$.

Fix $x \in X - N$. (4.4.5) and (4.4.7) imply that the stochastic process $(e^{-\alpha t} \tilde{u}_B(X_t), \mathscr{F}_t, P_x)$ is a right continuous non-negative supermartingale.

The optional sampling theorem [65; VI, T13] then gives $H_\alpha^B\tilde{u}_B(x) = E_x(e^{-\alpha\sigma_B}\tilde{u}_B(X_{\sigma_B})) \leqq \widetilde{u_B}(x)$. On the other hand, we have $H_\alpha^B\tilde{u}_B(x) = H_\alpha^B u(x)$ from (4.4.6). We get $H_\alpha^B u \leqq \tilde{u}_B$ q.e. proving (4.4.3). Q.E.D.

Proof of Theorem 4.4.1. Suppose first that $u \in \mathscr{F}$ is α-excessive with respect to $\{p_t, t > 0\}$. Then the reduced function u_B of u on B coincides with $P_{\mathscr{H}_\alpha^B} u$ as we saw in § 3.3 already. Hence, $H_\alpha^B u$ is a quasi-continuous version of $P_{\mathscr{H}_\alpha^B} u$ by virtue of Lemma 4.4.1.

In dealing with a general non-negative $u \in \mathscr{F}$, we may assume that u is bounded because, otherwise, the approximation of u by the truncations $u \wedge n$ works. For each $\beta > 0$, $R_\beta\tilde{u}$ can be then expressed as a difference of two bounded α-excessive functions. Hence, $H_\alpha^B(R_\beta\tilde{u})$ is a quasi-continuous version of $P_{\mathscr{H}_\alpha^B}(G_\beta u)$. On the other hand, since $\beta G_\beta u$ is $\mathscr{E}_1$-convergent to u as $\beta \to \infty$ on account of Lemma 1.3.3, so is $P_{\mathscr{H}_\alpha^B}(\beta G_\beta u)$ to $P_{\mathscr{H}_\alpha^B} u$. Furthermore, by choosing a properly exceptional set N such that the relation (4.3.4) holds for $\tilde{u}$ and $x \in X - N$, we can get $\lim_{\beta\to\infty} H_\alpha^B(\beta R_\beta\tilde{u})(x) = H_\alpha^B\tilde{u}(x)$, $x \in X - N$. Due to Theorem 3.1.4, we conclude that $H_\alpha^B\tilde{u}$ is a quasi-continuous version of $P_{\mathscr{H}_\alpha^B} u$. Q.E.D.

Define the kernel $R_\alpha^{X-B}(x, E)$, $x \in X$, $E \in \mathscr{B}(X)$, by

(4.4.8) $$R_\alpha^{X-B}(x, E) = E_x\left(\int_0^{\sigma_B} e^{-\alpha s} I_E(X_s)\, ds\right).$$

As we already saw in § 4.2, $\{R_\alpha^{X-B}, \alpha > 0\}$ is an m-symmetric resolvent kernel on $(X, \mathscr{B}^*(X))$ and the following Dynkin formula holds:

(4.4.9) $$R_\alpha f(x) = R_\alpha^{X-B} f(x) + H_\alpha^B R_\alpha f(x), \quad x \in X, \quad f \in \mathscr{B}^+(X).$$

For $f \in \mathscr{B}^+ \cap L^2(X; m)$, this formula expresses the orthogonal decomposition (4.4.2) of $R_\alpha f$ by virtue of Theorem 4.4.1 and Theorem 4.3.3. In particular,

Lemma 4.4.2. If $f \in \mathscr{B}^+ \cap L^2(X; m)$, then $R_\alpha^{X-B} f$ is quasi-continuous, $R_\alpha^{X-B} f \in \mathscr{F}_{X-B}$, and

(4.4.10) $$\mathscr{E}_\alpha(R_\alpha^{X-B} f, v) = (f, v), \quad v \in \mathscr{F}_{X-B}.$$

The restriction of the Dirichlet form $\mathscr{E}$ to $\mathscr{F}_{X-B}$ is denoted by $(\mathscr{F}_{X-B}, \mathscr{E})$. $\mathscr{F}_{X-B}$ is not necessarily dense in $L^2(X; m)$, but clearly

$(\mathscr{F}_{X-B}, \mathscr{E})$ is a Dirichlet form on $L^2(x, m)$ in the wide sense (see §1.4). According to Theorem 1.4.3, this is uniquely related to a Markovian resolvent on $L^2(X;m)$ which is not necessarily strongly continuous. Lemma 4.4.2 means that this resolvent is determined just by $\{R_\alpha^{X-B}, \alpha > 0\}$.

We further assume that the set $G = X - B$ is open. The set $\mathscr{F}_G$ is a subspace of $L_G^2(X; m) = \{u \in L^2(X; m): u = 0$ m-a.e. on $B\}$ which can be identified with $L^2(G; m)$ in an obvious manner. Hence, the form $(\mathscr{F}_G, \mathscr{E})$ can be considered as a Dirichlet form on $L^2(G; m)$ in the wide sense. We denote this by $\mathscr{E}_G$ and call it the *part of the Dirichlet form* $\mathscr{E}$ *on the open set* G. The part $\mathbf{M}_G$ of the Hunt process $\mathbf{M}$ on G has been introduced in §4.1 and, indeed, $\mathbf{M}_G$ is a standard process on $(G, \mathscr{B}(G))$ (Theorem 1.4.3) with an m-symmetric transition function on $(G, \mathscr{B}(G))$.

Theorem 4.4.2. Let G be an open set.
(i) The part $\mathscr{E}_G$ of the Dirichlet form $\mathscr{E}$ on G is a regular Dirichlet form on $L^2(G; m)$.
(ii) $\mathscr{E}_G$ is the Dirichlet form of the part of the process $\mathbf{M}$ on G, i.e., the strongly continuous semigroup $\{T_t^G, t > 0\}$ on $L^2(G; m)$ corresponding to $\mathscr{E}_G$ is determined by the transition function $\{p_t^G, t > 0\}$ of $\mathbf{M}_G$.
(iii) A subset of G is of capacity zero with respect to $\mathscr{E}_G$ if and only if it is of capacity zero with respect to $\mathscr{E}$. A function is quasi-continuous on G with respect to $\mathscr{E}_G$ if and only if it is quasi-continuous with respect to $\mathscr{E}$.
(iv) $p_t^G f$ is a quasi-continuous version of $T_t^G f$ for any $f \in \mathscr{B}^+(G) \cap L^2(G; m)$.

Proof. (i) On account of the regularity of $\mathscr{E}$ and Lemma 1.4.2 (ii), the space $\mathscr{F}_G \cap C_0(G)$ is uniformly dense in $C_0(G)$.

On the other hand, by taking $\mathscr{E}_\alpha$-orthogonal complements of the both hand sides of (3.3.17) in Theorem 3.3.4 (with $F = B$), we get

$$(4.4.11) \quad \overline{\mathscr{F}_G \cap C_0(G)} = \mathscr{F}_G,$$

where the left hand side denotes the $\mathscr{E}_{G,1}$-closure. Hence $\mathscr{E}_G$ is regular.
(ii) Since the resolvent of the process $\mathbf{M}_G$ is the restriction to $(G, \mathscr{B}(G))$ of $\{R_\alpha^G, \alpha > 0\}$ defined by (4.4.8), Lemma 4.4.2 reads as follows: $R_\alpha^G f \in \mathscr{F}_G$ and $\mathscr{E}_{G,\alpha}(R_\alpha^G f, v) = (f, v)_{L^2(G;m)}$, $\forall v \in \mathscr{F}_G$, for any $f \in \mathscr{B}^+(X) \cap L^2(G; m)$. Obviously, this implies the second assertion of the theorem.
(iii) Denote by $\mathrm{Cap}_G(A)$ the 1-capacity of an open set $A \subset G$ with respect to $\mathscr{E}_G$. Let $\{A_n\}$ be a sequence of decreasing open subsets of G. It is enough to show the equivalence $\mathrm{Cap}_G(A_n) \downarrow 0 \Leftrightarrow \mathrm{Cap}(A_n) \downarrow 0$.

The implication "$\Rightarrow$" is clear from the obvious inequality $\mathrm{Cap}_G(A) \geqq$

$\text{Cap}(A)$ for open $A \subset G$. Since $\mathscr{E}_G$ is the Dirichlet form of the (standard) process $\mathbf{M}_G$, the proof of Lemma 4.3.1 applies and we see that 1-equilibrium potential e_A^G with respect to $\mathscr{E}_G$ possesses as its version the (1-) hitting probability $p_A^G(x) = E_x(e^{-\sigma_A}; \sigma_A < \sigma_B)$ with respect to the process $\mathbf{M}_G$. If $\text{Cap}(A_n) \downarrow 0$, then $p_{A_n}(x) \downarrow 0$ m-a.e. and, consequently, $p_{A_n}^G(x) \downarrow 0$ m-a.e. Hence, $\text{Cap}_G(A_n) = \mathscr{E}_{G,1}(p_{A_n}^G, p_{A_n}^G) \to 0, n \to \infty$.
(iv) By (ii) and Lemma 4.4.2, $R_\alpha^G f$ is quasi-continuous on G with respect to $\mathscr{E}$ and hence with respect to $\mathscr{E}_G$ as well. The rest of the proof is the same as the proof of Theorem 4.3.3. Q.E.D.

Example 4.4.1. Let $\mathbf{M}$ be n-dimensional Brownian motion. According to Example 1.4.1, the Dirichlet form $\mathscr{E}$ of $\mathbf{M}$ is given by $(1/2\mathbf{D}, H^1(R^n))$. Let D be a domain in R^n. According to the identity (3.3.21) in Example 3.3.2, the part on D of the Dirichlet form $(1/2\ \mathbf{D},\ H^1(R^n))$ is nothing but the form $(1/2\ \mathbf{D},\ H_0^1(D))$ on $L^2(D)$.

By virtue of Theorem 4.4.2, the form $(1/2\mathbf{D}, H_0^1(D))$ is the Dirichlet form of the part of $\mathbf{M}$ on the domain D, which is often called the *absorbing barrier Brownian motion* on D. We can also conclude from (3.3.21) and Theorem 4.4.1 that a non-negative function $u \in H^1(R^n)$ is α-harmonic on D if and only if $\tilde{u} = H_\alpha^{R^n-D}\tilde{u}$.

Example 4.4.2. Consider the demi-space $D \subset R^n$, i.e., $D = \{x = (x', x_n) : x' \in R^{n-1}, x_n > 0\}$. We set $\partial D = \{x = (x', x_n) : x' \in R^{n-1}, x_n = 0\}$ and $\bar{D} = D \cup \partial D$. The Sobolev space $(1/2\mathbf{D}, H^1(D))$ is a Dirichlet space on $L^2(D)$ which is not regular (see Example 1.2.3). However if we regard this as a Dirichlet space on $L^2(\bar{D})$ rather than on $L^2(D)$, then this is regular because $H^1(D)$ densely contains the space $C_0^\infty(\bar{D})$ (the restrictions to $\bar{D}$ of functions in $C_0^\infty(R^n)$).

The associated diffusion $\hat{\mathbf{M}} = (\hat{X}_t, P_x)_{x \in \bar{D}}$ on $\bar{D}$ is the *reflecting barrier Brownian motion;* i.e., $\hat{\mathbf{M}}$ is obtained from the Brownian motion $\mathbf{M} = (X_t, P_x)_{x \in R^n}$ on R^n by setting

$$(4.4.12) \quad \hat{X}_t = (X_t', |X_t^{(n)}|), \quad \text{where } X_t = (X_t', X_t^{(n)}).$$

To verify this assertion, observe that the resolvent $\hat{R}_\alpha$ of $\hat{\mathbf{M}}$ is expressed by the resolvent R_α of $\mathbf{M}$ as follows:

$$(4.4.13) \quad \hat{R}_\alpha f(x) = R_\alpha f_1(x) + R_\alpha f_2(x), \quad x \in D, \quad f \in C_0^\infty(D).$$

Here f_1 is the extension of f by setting $f_1(x) = 0$, $x \in R^n - D$. f_2 is defined by $f_2(x) = 0$ for $x \in D$ and $f_2(x', x_n) = f(x', -x_n)$ for $x_n \leqq 0$. It is clear from this that $\hat{R}_\alpha f \in H^1(D)$, $(\alpha - (1/2)\Delta)\hat{R}_\alpha f(x) = f(x)$, $x \in D$, and $\lim_{x_n \downarrow 0} (\partial/\partial x_n)\hat{R}_\alpha f(x) = 0$. An integration by parts then gives

$$(4.4.14) \quad \frac{1}{2}\mathbf{D}(\hat{R}_\alpha f, v) + \alpha(\hat{R}_\alpha f, v) = (f, v), \quad v \in H^1(D),$$

where the domain of integration in each term is the demi-space D. Equation (4.4.14) implies that the Dirichlet space of $\hat{\mathbf{M}}$ (on $L^2(\bar{D})$) is just $(1/2\mathbf{D}, H^1(D))$.

Since $(1/2\mathbf{D}, H^1(D))$ is regular on $L^2(\bar{D})$, each function $u \in H^1(D)$ admits a modification $\tilde{u}$ on $\bar{D}$ which is quasi-continuous on $\bar{D}$, or, equivalently, fine continuous q.e. on $\bar{D}$ with respect to the reflecting barrier Brownian motion $\hat{\mathbf{M}}$. Then we can see in the same way as in Example 3.3.2 that

$$(4.4.15) \quad H_0^1(D) = \{u \in H^1(D);\ \ \tilde{u} = 0 \ \text{ q.e. on } \partial D\}.$$

Compare this with (1.2.15). We further see by Theorem 4.4.1 that $u \in H^1(D)$ is α-harmonic if and only if $\tilde{u} = H_\alpha^{\partial D}\tilde{u}$.

§4.5 Continuity, Killing, and Jumps of Sample Paths

Just as in the preceding two sections we consider the m-symmetric Hunt process $\mathbf{M}$ on X whose Dirichlet space $(\mathscr{F}, \mathscr{E})$ on $L^2(X; m)$ is regular. Recall that the Dirichlet space $(\mathscr{F}, \mathscr{E})$ is said to possess the local property if $\mathscr{E}(u, v)$ vanishes whenever $u, v \in \mathscr{F}$ have disjoint compact supports.

Lemma 4.5.1. The following two conditions are equivalent each other:

(i) $(\mathscr{F}, \mathscr{E})$ possesses the local property,

(ii) For each relatively compact open set A and $\alpha > 0$, the hitting distribution $H_\alpha^{X-A}(x, \cdot)$ is concentrated on the boundary ∂A for q.e. $x \in A$.

Proof. Denote by $\mathbf{P}$ the projection operator on the space $\mathscr{F}_A = \{u \in \mathscr{F}: \tilde{u} = 0 \text{ q.e. on } X - A\}$ in the Hilbert space $(\mathscr{F}, \mathscr{E}_\alpha)$. By virtue of Theorem 4.4.1, we have for non-negative $u \in \mathscr{F}$ that

$$(4.5.1) \quad \tilde{u} - H_\alpha^{X-A}\tilde{u} \text{ is a quasi-continuous version of } \mathbf{P}u.$$

Suppose that $\mathscr{E}$ possesses the local property. Take a non-negative $u \in \mathscr{F} \cap C_0(X)$ such that $\text{Supp}[u] \subset X - \bar{A}$. Since $\mathbf{P}v = v$ and $\text{Supp}[v] \subset \bar{A}$ for any $v \in \mathscr{F}_A$, we get $\mathscr{E}_\alpha(\mathbf{P}u, v) = \mathscr{E}_\alpha(u, v) = 0$, $\forall v \in \mathscr{F}_A$, which together with (4.5.1) implies $H_\alpha^{X-A}\tilde{u}(x) = 0$ q.e. on A. From this and Lemma 1.4.2. we conclude that $H_\alpha^{X-A}(x, X - \partial A) = 0$ q.e. $x \in A$.

Conversely, assume condition (ii) and take $u, v \in \mathscr{F}$ with disjoint compact supports. Without loss of generality, we may suppose that u is non-negative. Let A be a relatively compact open set such that $\text{Supp}[v] \subset A$, $\text{Supp}[u] \subset X - \bar{A}$. Then $\mathscr{E}(u, v) = \mathscr{E}_\alpha(u, v) = \mathscr{E}_\alpha(\mathbf{P}u, v)$ which vanishes because $\mathbf{P}u = 0$ in view of (4.5.1) and the assumption (ii). Q.E.D.

We call a Hunt process $\mathbf{M}$ on X a *diffusion* if

(4.5.2) $P_x(X_t \text{ is continuous in } t \in [0, \zeta)) = 1$

for every $x \in X$. At the end of § 4.3 we have defined the equivalence of two m-symmetric Hunt processes on X.

Theorem 4.5.1. The following conditions are equivalent each other:
(i) $(\mathscr{F}, \mathscr{E})$ possesses the local property,
(ii) **M** is of continuous sample paths for q.e. starting point, i.e., there exists a properly exceptional set N such that (4.5.2) holds for every $x \in X - N$,
(iii) There exists an m-symmetric diffusion on X which is equivalent to **M**.

Proof. Obviously, (ii) implies the second condition of Lemma 4.5.1. Furthermore, (ii) is equivalent to (iii) in view of Theorems 4.1.2 and 4.1.3. Hence, it only remains to prove the implication (i) $\Rightarrow$ (ii).

Assume the local property of $(\mathscr{F}, \mathscr{E})$ and let $\mathscr{O}$ be a countable open base of X. We may assume that each $A \in \mathscr{O}$ is relatively compact. By virtue of Lemma 4.5.1 and Theorem 4.2.1, there exists a properly exceptional set $N \subset X$ such that

(4.5.3) $H_\alpha^{X-A}(x, X - \partial A) = 0, \ {}^\forall x \in A - N,$

for every $A \in \mathscr{O}$, $\alpha > 0$ being fixed. We notice that the relation (4.5.3) then holds for every $\alpha > 0$ because of $H_\beta^{X-A}(x, X - \partial A) = H_\alpha^{X-A}(x, X - \partial A) + (\alpha - \beta) \int_{A-N} R_\beta^A(x, dy) H_\alpha^{X-A}(y, X - \partial A)$, $x \in A$.

Let us put $\Omega_d = \{\omega \in \Omega\colon {}^\exists t \in (0, \zeta(\omega)),\ X_{t-}(\omega) \neq X_t(\omega)\}$. On account of the sample path property (M. 6),
$\Omega_d = \bigcup_{A \in \mathscr{O}} \bigcup_{s \in Q+} \{\omega \in \Omega\colon 0 < \dot{\sigma}_{X-A}(\theta_s\omega) < \infty,\ X_{\dot{\sigma}_{X-A}(\theta_s\omega)}(\theta_s\omega) \in X - \partial A\}$.
For each $x \in X - N$, the P_x-measure of the event inside the braces equals $E_x(P_{X_s}(\sigma_{X-A} < \infty,\ X_{\sigma_{X-A}} \in X - \partial A);\ X_s \in A) = E_x(H_{0+}(X_s, X - \partial A);\ X_s \in A - N) = 0$. Hence, (4.5.2) holds. Q.E.D.

Having completely characterized the sample continuity up to the life time ζ by means of the local property of the Dirichlet form, our next task is to study the location $X_{\zeta-}(\in X_\Delta)$ of the sample path right before the life time ζ. When $X_{\zeta-} \in X$ with positive probability, we say that the *killing takes place inside X*.

In the following we occasionally denote the integral of a function v with a positive measure μ by $\langle \mu, v\rangle$ or $\langle v, \mu\rangle$:

$$(4.5.4)\quad \langle \mu, v\rangle = \langle v, \mu\rangle = \int_X v(x)\mu(dx).$$

Lemma 4.5.2. (i) There exists a unique positive Radon measure k on X charging no exceptional set such that for any $u \in \mathscr{F}$

$$(4.5.5)\quad \lim_{t \downarrow 0} \frac{1}{t}\int_X u(x)^2(1 - p_t 1(x))m(dx) = \langle \tilde{u}^2, k\rangle,$$

$\tilde{u}$ being any quasi-continuous version of u.
(ii) For any $f,\ h \in \mathscr{B}^+(X)$, and $t > 0$,

$$(4.5.6)\quad E_{h\cdot m}(f(X_{\zeta-}); \zeta \leq t) = \int_0^t \langle fk, p_s h\rangle ds.$$

(iii) $I_K \cdot k$ is of finite energy integral for each compact K. The function $E_x(e^{-\alpha\zeta}f(X_{\zeta-}))$ is a quasi-continuous version of the potential $U_\alpha(f\cdot k)$ for $\alpha > 0$ and $f \in C_0^+(X)$.

Proof. In view of the formula

$$(4.5.7)\quad \frac{1}{2t}\int_{X\times X}(u(x) - u(y))^2 p_t(x, dy)m(dx) +$$
$$\frac{1}{t}\int_X u(x)^2(1 - p_t 1(x))m(dx) = \frac{1}{t}(u, u - p_t u) \uparrow \mathscr{E}(u, u),\ t \downarrow 0,$$

holding for any Borel function $u \in \mathscr{F}$, we can subtract a sequence $t_n \downarrow 0$ such that the measure $(1/t_n)(1 - p_{t_n}1(x))m(dx)$ converges vaguely to a positive Radon measure k on X and

$$(4.5.8)\quad \langle \tilde{u}^2, k\rangle \leq \mathscr{E}(u, u),\ u \in \mathscr{F} \cap C_0(X).$$

In particular, $I_K \cdot k$ is of finite energy integral for any compact set K and, hence, k is smooth (see §3.2). k charges no exceptional set on account of Theorem 4.3.1.

Using Theorem 3.1.4, we easily see that the inequality (4.5.8) holds for any $u \in \mathscr{F}$. We can then show that the relation (4.5.5) holds for any $u \in \mathscr{F}$ provided that the limit on the left hand side is taken along the above sequence t_n. Indeed, choosing for any $\varepsilon > 0$ a function $v \in \mathscr{F} \cap C_0(X)$ such that $\sqrt{\mathscr{E}(u - v, u - v)} < \varepsilon$, it follows from (4.5.7) and (4.5.8) that

$$\left|\sqrt{\frac{1}{t_n}\langle u^2,(1-p_{t_n}1)\cdot m\rangle}-\sqrt{\langle \tilde{u}^2,k\rangle}\right|\leqq\left|\sqrt{\frac{1}{t_n}\langle v^2,(1-p_{t_n}1)\cdot m\rangle}\right.$$
$$\left.-\sqrt{\langle v^2,k\rangle}\right|+2\varepsilon$$

which is not greater than 3ε for sufficiently small t_n.

Next let us show the relation (4.5.6) for the measure k decided above. We first do this for $f\in C_0^+(X)$ and $h=\alpha R_\alpha \hat{h}$, $\hat{h}\in C_0^+(X)$. Notice that $h\in\mathscr{D}(A)\cap L^1(X;m)$, A being the infinitisimal generator of the L^2-semigroup T_t determined by p_t. We then have

$$E_{h\cdot m}(f(X_{\zeta-});\,\zeta\leqq t)=\lim_{n\to\infty}E_{h\cdot m}\left(\sum_{k=1}^{[t/t_n]}f(X_{(k-1)t_n})\cdot I_{\{(k-1)t_n<\zeta\leqq kt_n\}}\right)$$
$$=\lim_{n\to\infty}\sum_{k=1}^{[t/t_n]}E_{h\cdot m}[f(X_{(k-1)t_n})\{1-p_{t_n}1(X_{(k-1)t_n})\}]$$
$$=\lim_{n\to\infty}\sum_{k=1}^{[t/t_n]}(p_{(k-1)t_n}h,\ f(1-p_{t_n}1)),$$

t_n being a sequence tending to zero. Choose $\delta>0$ such that $\|p_sh-h\|_{L^2}\leqq 2s\|Ah\|_{L^2}$ for any $s<\delta$. For $t_n<\delta$

$$\left|\sum_{k=1}^{[t/t_n]}(p_{(k-1)t_n}h,f(1-p_{t_n}1))-\frac{1}{t_n}\int_0^t(p_sh,f(1-p_{t_n}1))ds\right|$$
$$\leqq\sum_{k=1}^{[t/t_n]}\frac{1}{t_n}\int_{(k-1)t_n}^{kt_n}|(p_{(k-1)t_n}h-p_sh,f(1-p_{t_n}1))|\,ds$$
$$+\frac{1}{t_n}\int_{t-[t/t_n]}^{t}|(p_sh,f(1-p_{t_n}1))|\,ds\leqq 2t\|Ah\|_{L^2}\cdot\|f(1-p_\delta 1)\|_{L^2}$$
$$+\|h\|_{L^2}\|f(1-p_\delta 1)\|_{L^2},$$

which decreases to zero as $\delta\downarrow 0$. Therefore, $E_{h\cdot m}(f(X_{\zeta-});\ \zeta\leqq t)=\lim_{n\to\infty}\frac{1}{t_n}\int_0^t(p_sh,f(1-p_{t_n}1))ds=\int_0^t\langle p_sh\cdot f,\ k\rangle ds$ proving (4.5.6) for the present f and h. Letting α tend to infinity and then using the monotone lemma, we know that (4.5.6) holds for any $f,h\in\mathscr{B}^+(X)$.

Taking the Laplace transform of both sides of (4.5.6), we readily see that the function $w(x)=E_x(e^{-\alpha\zeta}f(X_{\zeta-}))$ is a version of the potential $U_\alpha(f\cdot k)$ for $f\in C_0^+(X)$. Since w is α-excessive, it is quasi-continuous by Theorem 4.3.2, completing the proof of the third statement (iii). In particular, the Radon measure k is uniquely determined by the process **M** independently of the choice of $\{t_n\}$. Hence, the proof of (i) is also complete. Q.E.D.

We now identify the measure k in the above theorem with the killing measure (associated with the form $\mathscr{E}$) appearing in the Beurling-Deny formula.

Lemma 4.5.3. The measure k in Lemma 4.5.2 coincides with the killing measure in Theorem 2.2.1.

Proof. Denote by $\tilde{k}$ the killing measure in the statement of Theorem 2.2.1. In accordance with the construction of $\tilde{k}$ there, consider a sequence of relatively compact open sets G_l increasing to X. The difficulty lies in that $\tilde{k}$ was constructed in two steps and, in fact, $\tilde{k}$ is a vague limit of k_l determined by (2.2.9). However, a similar reasoning to the first part of the proof of Lemma 4.5.2 applies yielding that

$$\text{(4.5.9)} \quad \lim_{l\to\infty} \langle \tilde{u}^2, k_l\rangle = \langle \tilde{u}^2, \tilde{k}\rangle$$

for any $u \in \mathscr{F}$ with compact support.

Let $\mathscr{E}^{(l)}$ and $\mathbf{M}^{(l)}$ be the parts on the open set G_l of the Dirichlet form $\mathscr{E}$ and the Hunt process $\mathbf{M}$ respectively. By virtue of Theorem 4.4.2, we can apply Lemma 4.5.2 to $\mathbf{M}^{(l)}$ and $\mathscr{E}^{(l)}$ getting a positive Radon measure k_l^0 on G_l such that for any $u \in \mathscr{F}_{G_l}$

$$\text{(4.5.10)} \quad \lim_{t\downarrow 0} \frac{1}{t}\int_{G_l} u(x)^2(1 - p_t^{(l)}1(x))m(dx) = \int_{G_l} \tilde{u}(x)^2 k_l^0\,(dx)$$

and for any $f, h \in C_0^+(X)$ and sufficiently large l

$$\text{(4.5.11)} \quad E_{h.m}(f(X_{(\zeta\wedge\tau_l)-}); \zeta\wedge\tau_l \leq t) = \int_0^t \langle f\cdot k_l^0, p_s^{(l)}h\rangle ds,$$

where $\tau_l = \sigma_{X-G_l}$ and $p_t^{(l)}$ is the transition function of $\mathbf{M}^{(l)}$.

Comparing (4.5.10) with (2.2.9), we have $k_l^0(E) \geq k_l(E)$, $E \subset G_l$. Hence, the right hand side of (4.5.11) is not less than
$\int_0^t \langle f\cdot k_l, p_s^{(l)}h\rangle\, ds \geq \int_0^t \langle f\cdot k_l, p_s^{(l_0)}h\rangle ds,\ l > l_0.$
Since the left hand side of (4.5.11) tends to $E_{h\cdot m}(f(X_{\zeta-}); \zeta \leq t)$, we get from (4.5.6) and (4.5.9) the inequality

$$\frac{1}{t}\int_0^t \langle f\cdot k, p_s h\rangle ds \geq \frac{1}{t}\int_0^t \langle f\cdot \tilde{k}, p_s^{(l_0)}h\rangle\, ds.$$

After letting $l_0 \to \infty$ and then $t \downarrow 0$, we arrive at $\langle f\cdot k, h\rangle \geq \langle f\cdot \tilde{k}, h\rangle$. The converse inequality is obvious and, consequently, $k = \tilde{k}$. Q.E.D.

Finally, we are concerned with the jumping measure J appeared in the Beurling-Deny formula: J is the unique positive Radon measure on $X \times X - d$ such that for any $u, v \in \mathscr{F} \cap C_0(X)$ with disjoint support,

$$(4.5.12)\quad 2\int u(x)v(y)J(dxdy) = -\mathscr{E}(u, v).$$

Lemma 4.5.4. (i) J charges no part of $X \times X - d$ whose projection on the factor X *is* exceptional.
(ii) The formula (2.2.1) extends to any $u, v \in \mathscr{F}$:

$$(4.5.13)\quad \mathscr{E}(u, v) = \mathscr{E}^{(c)}(u, v) + \int_{X\times X-d} (\tilde{u}(x) - \tilde{u}(y))(\tilde{v}(x) - \tilde{v}(y))\, J(dxdy) + \int_X \tilde{u}(x)\tilde{v}(x)k(dx),$$

where $\tilde{u}$ and $\tilde{v}$ denote quasi-continuous versions and $\mathscr{E}^{(c)}$ is a symmetric form with $\mathscr{D}[\mathscr{E}^{(c)}] = \mathscr{F}$ satisfying for any $u, v \in \mathscr{F}$ of compact support

$$(4.5.14)\quad \mathscr{E}^{(c)}(u, v) = 0 \text{ if } v \text{ is constant on a neighbourhood of } \mathrm{Supp}[u].$$

Proof. Consider a relatively compact open set G and a non-negative $v \in \mathscr{F} \cap C_0(X)$ with $\mathrm{Supp}\,[v] \subset X - \bar{G}$. Then (4.5.12) implies that the measure $J_v(dx) = 2I_G(x) \int v(y)J(dxdy)$ is of finite energy integral with respect to the part $\mathscr{E}_G$ of the Dirichlet form $\mathscr{E}$ on the set G:

$$(4.5.15)\quad \int_G |u(x)|\, J_v(dx) \leqq C\sqrt{\mathscr{E}(u, u)},\ \forall u \in \mathscr{F}_G \cap C_0(G).$$

Hence, J_v charges no set of zero capacity with respect to $\mathscr{E}_G$ and, consequently, no exceptional set in view of Theorem 4.4.2.

This together with Lemma 4.5.2 implies that the integrals in (4.5.13) are independent of the choice of the versions $\tilde{u}$ and $\tilde{v}$. Let us define $\mathscr{E}^{(c)}$ by the equation (4.5.13). By virtue of Theorem 2.2.1, $\mathscr{E}^{(c)}$ then becomes a symmetric form. Its property (2.2.2) extends to (4.5.14) by virtue of Theorem 2.1.2 and the expression $v = (v - v_0) + v_0$ with $v_0 \in \mathscr{F} \cap C_0(X)$ such that $v_0 = v = \text{constant}$ on $\mathrm{Supp}[u]$. Q.E.D.

We can derive the following identity from formula (4.5.12) and Theorem 4.4.1: for G, v and J_v as in the above proof

$$(4.5.16)\quad v - H_\alpha^{X-G} v = -U_\alpha^G J_v,\quad \alpha > 0,$$

where $U_\alpha^G J_v$ is the α-potential of J_v with respect to the form $\mathscr{E}_G$. In particular, for any non-negative Borel h vanishing outside G

$$(4.5.17)\quad E_{h\cdot m}(e^{-\alpha\tau_G}v(X_{\tau_G})) = \int R_\alpha^G h(x)v(y)J(dxdy),$$

τ_G being the first leaving time from G: $\tau_G = \sigma_{X-G}$. This formula is strengthened in the next lemma.

Lemma 4.5.5. Let G be a relatively compact open set.
(i) For any $h, f, g \in \mathscr{B}_b^+(X)$ such that $\mathrm{Supp}[h] \subset G$, $\mathrm{Supp}[f] \subset G$ and $\mathrm{Supp}[g] \subset X - \bar{G}$,

$$(4.5.18)\quad E_{h\cdot m}(f(X_{\tau_G-})g(X_{\tau_G});\ \tau_G \leq t) = 2\int_0^t \left[\int p_s^G h(x)f(x)g(y)J(dxdy)\right] ds.$$

(ii) The function $E_x(e^{-\alpha\tau_G}f(X_{\tau_G-})g(X_{\tau_G}))$ on G is a quasi-continuous version of the potential $U_\alpha^G(fJ_g)$ for $\alpha > 0$ and $f, g \in C_0^+(X)$ with $\mathrm{Supp}[f] \subset G$, $\mathrm{Supp}[g] \subset X - \bar{G}$.

Proof. The proof proceeds in a similar way to the proof of Lemma 4.5.2 In fact, when h satisfies the condition in the proof of Lemma 4.5.2, we see that the left hand side of (4.5.18) equals $\lim_{\delta\downarrow 0} \sum_{k=1}^{[t/\delta]} (p_{(k-1)\delta}^G h,\ f\cdot p_\delta g) = \lim_{\delta\downarrow 0} \frac{1}{\delta}\int_0^t (p_s^G h,\ f\cdot p_\delta g)ds$, where we use the estimate $\|f\cdot p_\delta g\|_{L^2}^2 \leq (f^2, p_\delta g^2) \xrightarrow[\delta\downarrow 0]{} (f^2, g^2) = 0$.

The last limit is equal to the right hand side of (4.5.18) because the preceding lemma implies

$$2\int \tilde{u}(x)\tilde{v}(y)J(dxdy) = -\mathscr{E}(u, v) = \lim_{s\downarrow 0} \frac{1}{s}(u, p_s v)$$

for any $u, v \in \mathscr{F}$ with compact disjoint support. Q.E.D.

Summing up Lemma 4.5.2—Lemma 4.5.5, we can formulate a refinement of Theorem 2.2.1 as follows:

Theorem 4.5.2. The Dirichlet form $\mathscr{E}(u, v)$, $u, v \in \mathscr{F}$, admits a unique representation (4.5.13) with $\mathscr{E}^{(c)}$, J and k possessing the following properties:

The *diffusion form* $\mathscr{E}^{(c)}$ is a symmetric form with $\mathscr{D}[\mathscr{E}^{(c)}] = \mathscr{F}$ and has the property (4.5.14) which is stronger than the local property.

The *jumping measure* J is a unique positive Radon measure on $X \times X - d$ satisfying (4.5.12). J indicates the jumps of the sample paths in the sense of the identity (4.5.18).

The *killing measure* k is a unique positive Radon measure satisfying (4.5.5). k indicates the killing of the sample paths inside X in the sense of the identity (4.5.6).

Combining this theorem with Lemma 4.5.2 (iii) and Theorem 4.5.1, we get

Theorem 4.5.3. The following conditions are equivalent to each other:
(i) $\mathscr{E}$ possesses the property (4.5.14).
(ii) $\mathscr{E}$ possesses the local property and $k=0$.
(iii) **M** is of continuous sample paths and with no killing inside for q.e. starting point, i.e., there exists a properly exceptional set N such that (4.5.2) and

$$(4.5.19)\quad P_x(X_{\zeta-}\in X,\ \zeta<\infty)=0$$

hold for every $x\in X-N$.
(iv) **M** admits an equivalent m-symmetric diffusion on X satisfying (4.5.19) for every $x\in X$.

Let us define the *resurrected form* $\mathscr{E}^{\text{res}}$ on $L^2(X;\, m)$ by

$$(4.5.20)\quad \begin{cases} \mathscr{E}^{\text{res}}(u,v)=\mathscr{E}(u,v)-\int_X \tilde{u}(x)\tilde{v}(x)k(dx) \\ \mathscr{D}[\mathscr{E}^{\text{res}}]=\mathscr{F}. \end{cases}$$

By (4.5.13), $\mathscr{E}^{\text{res}}$ has the expression

$$(4.5.21)\quad \mathscr{E}^{\text{res}}(u,v)=\mathscr{E}^{(c)}(u,v)+\int(\tilde{u}(x)-\tilde{u}(y))(\tilde{v}(x)-\tilde{v}(y))J(dxdy),\quad u,v\in\mathscr{F}.$$

On the other hand, the relation (4.5.7) implies that every normal contraction operates on $\mathscr{E}^{\text{res}}$ and

$$(4.5.22)\quad \mathscr{E}^{\text{res}}(u,u)=\lim_{t\downarrow 0}\frac{1}{2t}E_m((u(X_t)-u(X_0))^2)$$

for any Borel $u\in\mathscr{F}$. This formula plays an important role in the next chapter.

Example 4.5.1. (one-dimensional diffusion).

Let $I = (r_1, r_2)$ be a one-dimensional interval and let m and k be positive Radon measures on I with $\mathrm{Supp}[m] = I$. We consider an m-symmetric Hunt process **M** on I whose Dirichlet space on $L^2(X; m)$ is given by $(\mathcal{F}^0, \mathcal{E})$ of Example 1.2.2.

As we saw in Example 3.3.3, no non-empty subset of I is of capacity zero. Accordingly any non-empty subset of I is neither exceptional nor polar with respect to the process **M** (Theorem 4.3.1).

From the expression (3.3.26) of the 1-equilibrium potential $p(x)$ of a one-point set $\{y\}$, we know that $p(x)$ is strictly positive. Hence Theorem 4.3.5 implies that

$$(4.5.23) \quad P_x(\sigma_y < \infty) > 0, \ {}^\forall x, y \in I,$$

namely, any two points have the communication with respect to the process **M**.

Since $\mathcal{E}$ possesses the local property, we see by Theorem 4.5.1 that **M** is a diffusion process on I with possible killing inside I indicated by the killing measure k. When $k = 0$ and $m =$ Lebesgue measure on I, then $(\mathcal{F}^0, \mathcal{E}) = (H_0^1(I), 1/2\mathbf{D})$ and, consequently, **M** is the absorbing barrier Brownian motion on the interval I in view of Example 4.4.1.

CHAPTER 5

Further Analysis by Additive Functionals

Just as in the last three sections, we assume throughout this chapter that we are given a symmetric Hunt process whose Dirichlet space is regular.

It is shown in §5.1 that all equivalence classes of PCAF's of the Hunt process are in one-to-one correspondence with all smooth measures introduced in §3.2. Such a correspondence is useful in that it enables us to reduce the calculus of functionals to the calculus of measures.

Sections 5.2, 5.3, and 5.4 are concerned with additive functionals which are not necessarily non-negative. In particular, the AF $A_t^{[u]} = \tilde{u}(X_t) - \tilde{u}(X_0)$ generated by the function u in the Dirichlet space admits a unique decomposition

$$A^{[u]} = M^{[u]} + N^{[u]},$$

where $M^{[u]}$ is a martingale AF of finite energy and $N^{[u]}$ is a continuous AF of zero energy (§5.2). $N_t^{[u]}$ is not of bounded variation in t unless u is a potential of a signed smooth measure, in a certain sense (§5.3). Thus, the decomposition in §5.2 may be regarded as a generalization (toward the Dirichlet space) of the Doob-Meyer decomposition of supermartingales and Ito's formula on semi-martingales.

A notion of stochastic integrals and a generalized Ito's formula for the MAF $M^{[u]}$ are formulated in §5.4, which are then used to derive the Beurling-Deny second formula.

A time changed process of the Hunt process by means of a PCAF gives rise to a Dirichlet space (the time changed Dirichlet space) living on the support of the AF. A rule relating the time changed Dirichlet space to the original one is established in §5.5.

Each section contains examples concerning the Brownian motion or more general diffusions.

§5.1 Positive Continuous Additive Functionals and Smooth Measures

Let (X, m) be as in §1.1 and **M** be as in §4.3, namely, $\mathbf{M} = (\Omega, \mathscr{M}, X_t,$

P_x) is an m-symmetric Hunt process on X whose Dirichlet space $(\mathscr{F}, \mathscr{E})$ on $L^2(X; m)$ is regular. Potential theories relevent to $(\mathscr{F}, \mathscr{E})$ and $\mathbf{M}$ have been developed in the preceding two chapters. We now utilize these theories and study the structures of some important classes of additive functionals of the Hunt process $\mathbf{M}$.[1]

Let us call an extended real valued function $A_t(\omega)$, $t \geqq 0$, $\omega \in \Omega$, an *additive functional* (*AF* in abbreviation) if $A_t(\omega)$ is a perfect additive functional in the ordinary sense[2] but with respect to the restricted Hunt process $\mathbf{M}|_{X-N}$, N being a properly exceptional set which depends on A in general.

More precisely, in order for $A_t(\omega)$ to be an AF the following conditions must hold:

(A.1) $A_t(\cdot)$ is $\mathscr{F}_t$-measurable, $\{\mathscr{F}_t\}$ being the minimum completed admissible family (see §4.1);

(A.2) there exist a set $\Lambda \in \mathscr{F}_\infty$ and an exceptional set $N \subset X$ such that $P_x(\Lambda) = 1$, $^\forall x \in X - N$, $\theta_t \Lambda \subset \Lambda$, $^\forall t > 0$, and moreover, for each $\omega \in \Lambda$, $A_\cdot(\omega)$ is right continuous and has the left limit on $[0, \infty]$, $A_0(\omega) = 0$, $|A_t(\omega)| < \infty$, $^\forall t < \zeta(\omega)$, $A_t(\omega) = A_{\zeta(\omega)}(\omega)$, $^\forall t \geqq \zeta(\omega)$ and $A_{t+s}(\omega) = A_s(\omega) + A_t(\theta_s\omega)$, $^\forall t, s \geqq 0$.

The sets Λ and N referred to in the preceding paragraph are called a *defining set* and an *exceptional set* of the AF $A_t(\omega)$ respectively.

Two AF's $A^{(1)}$ and $A^{(2)}$ are said to be *equivalent* if for each $t > 0$ $P_x(A_t^{(1)} = A_t^{(2)}) = 1$ q.e. $x \in X$. By virtue of Theorem 4.2.1, we can then find a common defining set Λ and a common properly exceptional set N of $A^{(1)}$ and $A^{(2)}$ such that $A_t^{(1)}(\omega) = A_t^{(2)}(\omega)$, $^\forall t > 0$, $^\forall \omega \in \Lambda$.

An AF which is non-negative and continuous on its defining set is called a *positive continuous AF* (*PCAF* in abbreviation). The set of all PCAF's is denoted by $\mathbf{A}_c^+$. The aim of this section is to characterize $\mathbf{A}_c^+$ by the class S of smooth measures associated with $(\mathscr{F}, \mathscr{E})$ (see §3.2). To this end we first construct a PCAF for any $\mu \in S_0$–the positive Radon measure of finite energy integral.

Lemma 5.1.1. For any $u \in \mathscr{F}$, $\nu \in S_0$, $0 < T < \infty$ and $\varepsilon > 0$,

$$(5.1.1)\quad P_\nu(\sup_{0 \leqq t \leqq T} |\tilde{u}(X_t)| > \varepsilon) \leqq \frac{e^T}{\varepsilon} \sqrt{\mathscr{E}_1(\nu)}\ \sqrt{\mathscr{E}_1(u, u)}.$$

Proof. Taking a quasi-continuous Borel version $\tilde{u}$, we let $E = \{x \in X\colon |\tilde{u}(x)| > \varepsilon\}$. Then the left hand side of (5.1.1) is dominated by $e^T \int_X p(x)\nu(dx)$ with $p(x) = E_x(e^{-\sigma_E})$. By virtue of Theorem 3.3.1 and Theorem 4.3.5, we have

1) The convention made in the last chapter is still in force in this chapter: any numerical function f on X is extended to X_Δ by setting $f(\Delta) = 0$.

2) Cf. R.M.Blumenthal-R.K.Getoor [14].

$$\int_X p(x)\nu(dx) = \mathscr{E}_1(p, U_1\nu) \leqq \sqrt{\mathscr{E}_1(\nu)}\sqrt{\mathrm{Cap}(E)}$$

$$\leqq \frac{1}{\varepsilon}\sqrt{\mathscr{E}_1(\nu)}\sqrt{\mathscr{E}_1(u, u)}.\quad \text{Q.E.D.}$$

Lemma 5.1.2. Let $\{u_n\}$ be a sequence of quasi-continuous functions in $\mathscr{F}$. If $\{u_n\}$ is an $\mathscr{E}_1$-Cauchy sequence, then there exists a subsequence $\{u_{n_k}\}$ satisfying the condition that for q.e. $x \in X$. $P_x(u_{n_k}(X_t)$ converges uniformly in t on each compact interval of $[0, \infty)) = 1$.

Proof. Take $\{n_k\}$ such that $\sqrt{\mathscr{E}_1(u_{n_{k+1}} - u_{n_k}, u_{n_{k+1}} - u_{n_k})} \leqq 2^{-2k}$. By virtue of the preceding lemma, $P_\nu(\Lambda_k) \leqq e^T 2^{-k}\sqrt{\mathscr{E}_1(\nu)}$ with $\Lambda_k = \{\sup_{0 \leqq t \leqq T} |u_{n_k}(X_t) - u_{n_{k+1}}(X_t)| > 2^{-k}\}$. By the Borel-Cantelli lemma, the P_ν-measure of the set $\Lambda = \overline{\lim_k}\, \Lambda_k$ vanishes for every $\nu \in S_0$. Hence, $P_x(\Lambda) = 0$ q.e. by virtue of Theorem 3.3.2. Q.E.D.

Henceforth we use the notions $\langle \nu, u\rangle = \langle u, \nu\rangle = \int_X u(x)\nu(dx)$ for a measure ν and a function u.

Theorem 5.1.1. For any $\mu \in S_0$, there exists $A \in \mathbf{A}_c^+$ such that $E_x(\int_0^\infty e^{-t}dA_t)$ is a quasi-continuous version of $U_1\mu$.

Proof. Take a non-negative finite Borel and quasi-continuous version u of $U_1\mu$ such that

$$(5.1.2)\quad \begin{cases} nR_{n+1}u(x) \uparrow u(x), & n \to \infty, \forall x \in X - N, \\ u(x) = 0, \forall x \in N, \end{cases}$$

where N is some properly exceptional set and $\{R_\alpha, \alpha > 0\}$ is the resolvent of the process $\mathbf{M}$. We let

$$g_n(x) = \begin{cases} n(u(x) - nR_{n+1}u(x)), & x \in X - N \\ 0 \qquad\qquad , & x \in N. \end{cases}$$

Then it is easy to see that $g_n \cdot m \to \mu$ vaguely, $R_1 g_n(x) \uparrow u(x)$, $\forall x \in X - N$, and, moreover, $R_1 g_n$ is $\mathscr{E}_1$-convergent to u.

We define the approximating 1-order PCAF $\tilde{A}_n$ of $\mathbf{M}$ by $\tilde{A}_n(t, \omega) = \int_0^t e^{-s} g_n(X_s(\omega))ds$. Then, for any $\nu \in S_{00}$

$$(5.1.3)\quad E_\nu((\tilde{A}_n(\infty) - \tilde{A}_l(\infty))^2) \leqq 2\, M_\nu \sqrt{\mathscr{E}_1(\mu)}\, \|R_1 g_n - R_1 g_l\|_{\mathscr{E}_1}{}^{1)}$$

1) $\|v\|_{\mathscr{E}_1} = \sqrt{\mathscr{E}_1(v, v)}$.

with $M_\nu = \|U_2\nu\|_\infty$. In fact, by setting $g_{n,l} = g_n - g_l$, $n > l$, the left hand side of (5.1.3) is equal to

$$2E_\nu\left(\int_0^\infty e^{-s}g_{n,l}(X_s)ds \int_s^\infty e^{-u}g_{n,l}(X_u)du\right) = 2E_\nu\left(\int_0^\infty e^{-2s}g_{n,l}(X_s)R_1g_{n,l}(X_s)ds\right)$$
$$= 2\langle \nu, R_2(g_{n,l}R_1g_{n,l})\rangle = 2(U_2\nu, g_{n,l}R_1g_{n,l}) \leqq 2(U_2\nu, g_nR_1g_{n,l})$$
$$\leqq 2M_\nu(g_n, R_1g_n - R_1g_l) = 2M_\nu\mathscr{E}_1(R_1g_n, R_1g_n - R_1g_l).$$

Applying Schwarz inequality and noting $\mathscr{E}_1(R_1g_n, R_1g_n) = (g_n, R_1g_n) \leqq (g_n, u) = \langle \mu, R_1g_n\rangle \leqq \langle \mu, u\rangle = \mathscr{E}_1(\mu)$, we obtain (5.1.3).

Since $E_\nu(\tilde{A}_n(\infty)\,|\,\mathscr{F}_t) = \tilde{A}_n(t) + e^{-t}E_{X_t}(\tilde{A}_n(\infty)) = \tilde{A}_n(t) + e^{-t}R_1g_n(X_t)$, we see that

$$(5.1.4)\quad M_n(t) = \tilde{A}_n(t) + e^{-t}R_1g_n(X_t), \quad 0 \leqq t \leqq \infty,$$

is a martingale with respect to $(\mathscr{F}_t, P_\nu)$, $\nu \in S_{00}$. By Doob's inequality[1)]

$$P_\nu(\sup_{0\leqq t\leqq\infty} |M_n(t) - M_l(t)| > \varepsilon) \leqq \frac{1}{\varepsilon^2} E_\nu((\tilde{A}_n(\infty) - \tilde{A}_l(\infty))^2)$$

which, combined with (5.1.3), means that there eixsts a subsequence $\{n_k\}$ such that

$$(5.1.5)\quad P_\nu(M_{n_k}(t) \text{ converges uniformly on } [0, \infty)) = 1,\ \forall\nu \in S_{00}.$$

In view of (5.1.4), (5.1.5), Theorem 3.3.2 and Lemma 5.1.2, we conclude that, by selecting a new subsequence if necessary, $P_x(\Lambda) = 1$, $x \in X - \tilde{N}$, where

$$(5.1.6)\quad \Lambda = \{\omega \in \tilde{\Omega}:\ \tilde{A}_{n_k}(\infty, \omega) < \infty,\ \tilde{A}_{n_k}(t, \omega) \text{ converges uniformly in } t \text{ on each finite interval of } [0, \infty)\}$$ [2)]

and $\tilde{N}$ is some properly exceptional set containing N.

Let us set $\tilde{A}(t, \omega) = \lim_{n_k\to\infty} \tilde{A}_{n_k}(t, \omega)$ for $\omega \in \Lambda$ and $\tilde{A}(t, \omega) = 0$ for $\omega \notin \Lambda$. Furthermore, let $A(t,\omega) = \int_0^t e^s d\tilde{A}(s,\omega)$. A is then a PCAF with Λ and $\tilde{N}$ being its defining set and exceptional set respectively.

In order to complete the proof, it suffices to show $E_\nu(\tilde{A}(\infty)) = \langle \nu, u\rangle$,

1) Cf. P.A.Meyer [65].
2) $\tilde{\Omega}$ is the set defined by (4.1.6) for $\tilde{S} = X - \tilde{N}$.

$\forall\nu \in S_{00}$, by virtue of Theorem 3.3.2. Since $M_n(\infty) = \tilde{A}_n(\infty)$ is $L^2(P_\nu)$-convergent, so is the martingale $M_n(t)$. Hence, $E_\nu(\tilde{A}(t)) + e^{-t}\langle\nu, p_t u\rangle = \lim_{n\to\infty} E_\nu(M_n(t)) = \lim_{n\to\infty} E_\nu(\tilde{A}_n(\infty)) = \lim_{n\to\infty}\langle\nu, R_1 g_n\rangle = \langle\nu, u\rangle$. By letting t tend to infinity and noting the bound $\langle\nu, p_t u\rangle = \mathscr{E}_1(U_1\nu, p_t u) \leq \sqrt{\mathscr{E}_1(\nu)}\sqrt{\mathscr{E}_1(u, u)}$, we get the desired equality. Q.E.D.

Lemma 5.1.3 Consider μ and A of the preceding theorem. Then for any $\alpha > 0$ and bounded non-negative Borel function f, $E_x(\int_0^\infty e^{-\alpha t} f(X_t)dA_t)$ is a quasi-continuous version of $U_\alpha(f\cdot\mu)$.

Proof. It is sufficient to consider the case where $\alpha = 1$ and $f = I_G$, G being an open set with $\mu(\partial G) = 0$. Let $\phi(x) = E_x(\int_0^\infty e^{-t} I_G(X_t)dA_t)$, $\psi(x) = E_x(\int_0^\infty e^{-t} I_{X-G}(X_t)dA_t)$; then both ϕ and ψ are 1-excessive and $\phi + \psi = U_1\mu$. By Theorem 3.2.1 and Lemma 3.3.2, ϕ and ψ are quasi-continuous versions of 1-potentials of some measures λ and $\nu \in S_0$ respectively.

We know from Lemma 4.4.1 and the equalities $\phi = H_1^G\phi$, $\psi = H_1^{X-G}\psi$ that $\text{Supp}[\lambda] \subset \bar{G}$ and $\text{Supp}[\nu] \subset X - G$. Since $\mu = \lambda + \nu$, we have $\lambda = I_G\cdot\mu$. Q.E.D.

Theorem 5.1.2. For $\mu \in S_0$, $A \in \mathbf{A}_c^+$ of Theorem 5.1.1 is unique (up to equivalence).

Proof. Suppose that $A^{(1)}$, $A^{(2)} \in \mathbf{A}_c^+$ are associated with $\mu \in S_0$ in the manner of Theorem 5.1.1. Then $E_x(\int_0^\infty e^{-t}dA_t^{(1)}) = E_x(\int_0^\infty e^{-t}dA_t^{(2)}) = u(x)$, $x \in X - N$, for some properly exceptional set N. In the same way as in the proof of Theorem 5.1.1,

$$v_{ij}(x) = E_x\left(\int_0^\infty e^{-t}dA_t^{(i)}\int_0^\infty e^{-t}dA_t^{(j)}\right) = \sum_{k=i,j} E_x\left(\int_0^\infty e^{-2t}u(X_t)dA_t^{(k)}\right),$$

$x \in X - N$, $i,j = 1,2$. Let $u_n = u \wedge n$. Then, by Lemma 5.1.3, $\langle\nu, v_{ij}\rangle$

$$= \lim_{n\to\infty}\sum_{k=i,j} E_\nu\left(\int_0^\infty e^{-2t}u_n(X_t)dA_t^{(k)}\right) = \lim_{n\to\infty} 2\langle\nu, U_2(u_n\mu)\rangle = 2\langle U_2\nu\cdot u, \mu\rangle < \infty$$

for any $\nu \in S_{00}$. Hence, $E_\nu\left(\left\{\int_0^\infty e^{-t}dA_t^{(1)} - \int_0^\infty e^{-t}dA_t^{(2)}\right\}^2\right) = \langle\nu, v_{11} - 2v_{12} + v_{22}\rangle = 0$, $\nu \in S_{00}$, from which it easily follows that $A^{(1)}$ is equivalent to $A^{(2)}$. Q.E.D.

Henceforth we use the notations $(fA)_t = \int_0^t f(X_s)dA_s$, $U_A^\alpha f(x) = E_x\left(\int_0^\infty e^{-\alpha t} f(X_t)dA_t\right)$ for $A \in \mathbf{A}_c^+$ and $f \in \mathscr{B}^+$.

Lemma 5.1.4. For $\mu \in S_0$ and $A \in \mathbf{A}_c^+$, the following conditions are equivalent to each other:
(i) $U_A^1 1$ is a quasi-continuous version of $U_1\mu$,
(ii) $(h, U_A^\alpha f) = \langle f\cdot\mu, R_\alpha h\rangle$, $\alpha > 0$, $f, h \in \mathscr{B}^+$,
(iii) $E_{h\cdot m}((fA)_t) = \int_0^t \langle f\cdot\mu, p_s h\rangle ds$, $t > 0$, $f, h \in \mathscr{B}^+$,
(iv) $\lim_{t\downarrow 0} (1/t) E_{h\cdot m}((fA)_t) = \langle f\cdot\mu, h\rangle$ for any γ-excessive function h ($\gamma \geqq 0$) and $f \in \mathscr{B}^+$,
(v) $\lim_{t\downarrow 0} (1/t) E_{h\cdot m}(A_t) = \langle\mu, \tilde{h}\rangle$, $h \in \mathscr{B}^+ \cap \mathscr{F}$,
vi) $\alpha(h, U_A^{\alpha+\gamma} f) \uparrow \langle f\cdot\mu, h\rangle$, $\alpha \uparrow \infty$, for any γ-excessive function h ($\gamma \geqq 0$ and $f \in \mathscr{B}^+$,
(vii) $\lim_{\alpha\to\infty} \alpha(h, U_A^\alpha 1) = \langle\mu, \tilde{h}\rangle$, $h \in \mathscr{B}^+ \cap \mathscr{F}$.

Proof. (i) is equivalent to (ii) by Lemma 5.1.3. We can also see the equivalence of (ii) and (iii) by the uniqueness of the Laplace transform. The implications (iii) ⇒ (iv) ⇒ (vi) ⇒ (ii), (iii) ⇒ (v) and (ii) ⇒ (vii) are clear. Suppose that (v) is satisfied and put $c_t(x) = E_x(A_t)$. Then

$$\int_0^t \langle\mu, p_s h\rangle ds = \lim_{u\downarrow 0} \frac{1}{u}\int_0^t (p_s h, c_u)\, ds = \lim_{u\downarrow 0} \frac{1}{u}\int_0^t (h, c_{u+s} - c_s)\, ds$$

$$= \lim_{u\downarrow 0} \frac{1}{u}\int_t^{t+u} (h, c_s) ds - \lim_{u\downarrow 0} \frac{1}{u}\int_0^u (h, c_s) ds = (h, c_t)$$

for any γ-excessive function h in $\mathscr{F}$. By taking the Laplace transform we get (ii) for h of this type and $f = 1$, which is enough to obtain (i). In the same way we can derive the implication (vii) ⇒ (i). Q.E.D.

Lemma 5.1.5. Let $\mu \in S_0$ and $A \in \mathbf{A}_c^+$ be related as in Lemma 5.1.4. Then for any closed set $F \subset X$

$$\alpha E_{h\cdot m}\left(\int_0^{\sigma_{X-F}} e^{-(\alpha+\gamma)t} f(X_t) dA_t\right) \uparrow \int_{F^{(0)}} h(x) f(x) \mu(dx), \quad \alpha \uparrow \infty,$$

where h is any γ-excessive function ($\gamma \geqq 0$), $f \in \mathscr{B}^+$ and $F^{(0)}$ is the fine interior of F, i.e., $F^{(0)} = \{x \in X : \mathbf{P}_x(\sigma_{X-F} > 0) = 1\}$ $(\subset F)$.

Proof. It suffices to give the proof when h is bounded excessive and belongs to $L^2(X; m)$ and $f \in \mathscr{B}^+$ is bounded. By Lemma 5.1.3, $U_A^\alpha f$ is a quasi-continuous version of $U_\alpha(f\mu)$. By virtue of Theorem 4.4.1,

$$\alpha E_{h\cdot m}\left(\int_0^{\sigma_{X-F}} e^{-\alpha t} f(X_t) dA_t\right) = \alpha(h, U_A^\alpha f - H_\alpha^{X-F} U_A^\alpha f)$$

$$= \alpha\mathscr{E}_\alpha(R_\alpha h, U_A^\alpha f - H_\alpha^{X-F} U_A^\alpha f) = \alpha\mathscr{E}_\alpha(R_\alpha h - H_\alpha^{X-F} R_\alpha h, U_A^\alpha f)$$

$$= \alpha\mathscr{E}_\alpha(R_\alpha^F h, U_\alpha(f\mu)) = \alpha \int_X R_\alpha^F h(x) f(x)\mu(dx)$$

which increases to $\int_{F(0)} h(x)f(x)\mu(dx)$ as $\alpha \uparrow \infty$. Here $R_\alpha^F h(x) = E_X\left(\int_0^{\sigma_{X-F}} e^{-\alpha t} h(X_t)dt\right)$. Q.E.D.

Before proceeding to the main theorem of this section, we need one more lemma. Recall the definition of the smooth measure given in §3.2: $\mu \in S$ if μ is a positive Borel measure charging no set of zero capacity and there exists an increasing sequence $\{F_n\}$ of compact sets satisfying conditions (3.2.10), (3.2.11) and (3.2.12). Thus the class S contains all positive Radon measures charging no set of zero capacity.

Lemma 5.1.6. Let $\{F_n\}$ be an increasing sequence of closed sets. $\{F_n\}$ satisfies (3.2.12) if and only if

(5.1.7) $P_x(\lim_{n\to\infty} \sigma_{X-F_n} < \zeta) = 0$ q.e. $x \in X$.

Proof. Condition (3.2.12) is equivalent to

(5.1.8) $P_x(\lim_{n\to\infty} \sigma_{G-F_n} = \infty) = 1$ q.e.

for any relatively compact open set G on account of (4.3.5). (5.1.7) implies (5.1.8) because of the quasi-left continuity of the Hunt process **M**. To get (5.1.7) from (5.1.8), it suffices to choose a sequence $\{G_l\}$ of relatively compact open sets with $\bar{G}_l \subset G_{l+1}$, $G_l \uparrow X$, and observe the inequality $\sigma_{X-F_n} \geqq \sigma_{G_l-F_n} \wedge \sigma_{X-G_l}$. Q.E.D.

Theorem 5.1.3. The family of all equivalence classes of $\mathbf{A}_c^+$ and the family S are in one to one correspondence specified by the following relation:

(5.1.9) $\lim_{t\downarrow 0} \frac{1}{t} E_{h\cdot m}((f\cdot A)_t) = \langle f\cdot\mu, h\rangle$, $A \in \mathbf{A}_c^+$, $\mu \in S$,

for any γ-excessive function $h(\gamma \geqq 0)$ and $f \in \mathscr{B}^+$.

The proof of Theorem 5.1.3 is accomplished by the following two lemmas.

Lemma 5.1.7. Given $A \in \mathbf{A}_c^+$, there exists a unique $\mu \in S$ such that (5.1.9) holds.

Proof. For a given $A \in \mathbf{A}_c^+$ with a properly exceptional set $N \subset X$, we put $\phi(x) = E_x(\int_0^\infty e^{-t} f(X_t) e^{-A_t} dt)$, $x \in X - N$, where f is a Borel function in $L^2(X; m)$ such that $f(x) > 0$, $\forall x \in X$. Then, $\phi(x) > 0$, $\forall x \in X - N$. It is easy to see that the difference $R_1 f - \phi$ is 1-excessive on $X - N$. In particular, the function ϕ is, together with $R_1 f$, finely continuous on $X - N$, and consequently, right continuous along sample paths. This enables us to obtain the following expression of the difference:

$$(5.1.10) \quad U_A^1 \phi(x) = R_1 f(x) - \phi(x), \; x \in X - N.$$

Lemma 3.3.2 and Theorem 4.3.2 now imply that $U_A^1 \phi$ is, together with $R_1 f$, a quasi-continuous function in $\mathscr{F}$ and so is ϕ. Accordingly there exists a sequence $\{E_n\}$ of increasing closed sets such that $\mathrm{Cap}(X - E_n) \downarrow 0$, $n \to \infty$, $N \subset \bigcap_{n=1}^{\infty} (X - E_n)$ and $\phi|_{E_n}$ is continuous for each n.

Let us put

$$(5.1.11) \quad F_n = \left\{ x \in E_n \; : \phi(x) \geqq \frac{1}{n} \right\}$$

and prove that $\{F_n\}$ satisfies condition (5.1.7). To this end, set $B_n = \{x \in X - N\colon \phi(x) \leqq \frac{1}{n}\}$, $\sigma_n = \sigma_{B_n}$ and $\sigma = \lim_{n\to\infty} \sigma_n$. Since ϕ is finely continuous on $X - N$, we have for $x \in X - N$, $E_x(\int_{\sigma_n}^\infty e^{-t} f(X_t) e^{-A_t} dt) = E_x(e^{-\sigma_n} e^{-A_{\sigma_n}} \phi(X_{\sigma_n})) \leqq \frac{1}{n}$. By letting n tend to infinity, we can see $P_x(\sigma < \zeta) = 0$, $x \in X - N$, in view of the strict positivity of f. Hence $\{F_n\}$ satisfies (5.1.7) because of the inclusion $X - F_n \subset (X - E_n) \cup B_n$ and (4.3.5).

Now let $A_n = \mathrm{I}_{F_n} \cdot A$. By virtue of the inequality $U_{A_n}^1 1 \leqq n U_A^1 \phi (\in \mathscr{F})$ and Lemma 3.3.2 again, there exists a unique $\mu_n \in S_0$ such that $U_{A_n}^1 1$ is a quasi-continuous version of the potential $U_1 \mu_n$. But then

$$(5.1.12) \quad \mu_n = \mathrm{I}_{F_n} \cdot \mu_l, \; n < l$$

because $U_{A_n}^1 1 = U_{A_l}^1 I_{F_n}$ is a version of $U_1 I_{F_n} \cdot \mu_l$ by Lemma 5.1.3. We can now define a measure μ by $\mathrm{I}_{F_n} \cdot \mu = I_{F_n} \cdot \mu_n$, $n = 1, 2, \ldots$, $\mu(X - \bigcup_n F_n) = 0$. μ is smooth in view of Theorem 3.2.3 and Lemma 5.1.6.

It remains to show that A and μ are related by (5.1.9). By Lemma 5.1.4, we see for any $f \in \mathscr{B}^+$ and γ-excessive function h, $\langle f\mu, h \rangle =$

$\lim_{n\to\infty}\lim_{\alpha\to\infty} \alpha(h, U_{A_n}^{\alpha+\gamma} f) = \lim_{\alpha\to\infty}\lim_{n\to\infty} \alpha(h, U_A^{\alpha+\gamma}(I_{F_n}\cdot f))$. By (5.1.7), we get $\langle f\mu, h\rangle = \lim_{\alpha\to\infty} \alpha(h, U_A^\alpha f)$, which is obviously equivalent to (5.1.9). Q.E.D.

Lemma 5.1.8. Given $\mu \in S$, there exists $A \in \mathbf{A}_c^+$ uniquely up to equivalence such that (5.1.9) holds.

Proof. By Theorem 3.2.3 and Lemma 5.1.6, there exists a sequence $\{F_n\}$ of increasing closed sets satisfying (3.2.11) and (5.1.7) and $I_{F_n}\cdot\mu \in S_0$ for each n. By Theorem 5.1.1, Theorem 5.1.2 and Lemma 5.1.4, there exists $A^{(n)} \in \mathbf{A}_c^+$ uniquely up to the equivalence such that $A^{(n)}$ is related to $I_{F_n}\cdot\mu$ by (5.1.9). But then

(5.1.13) $A^{(n)} \sim I_{F_n}\cdot A^{(l)},\quad n < l$

because $I_{F_n}\cdot A^{(l)}$ is related by (5.1.9) to $I_{F_n}\cdot I_{F_l}\cdot\mu = I_{F_n}\cdot\mu$.

Choose a properly exceptional set N and a defining set Λ which are common to all $A^{(n)}$ such that, for any $\omega \in \Lambda$, $A_t^{(n)}(\omega) = (I_{F_n}\cdot A^{(n+1)})_t(\omega)$, $\forall t > 0$, $n = 1, 2, \cdots$, and $\sigma(\omega)$ $(= \lim_{n\to\infty} \sigma_{X-F_n}(\omega)) \geq \zeta(\omega)$. For $\omega \in \Lambda$ let

$$\begin{cases} A_t(\omega) = A_t^{(n)}(\omega), & \sigma_{X-F_{n-1}}(\omega) \leq t < \sigma_{X-F_n}(\omega), \quad n = 1, 2, \cdots \\ A_t(\omega) = A_{\sigma(\omega)-}(\omega), & t \geq \sigma(\omega). \end{cases}$$

Obviously $A \in \mathbf{A}_c^+$. Since $A_t = A_t^{(n)}$ for $t < \sigma_{X-F_n}$, we see from Lemma 5.1.5, $\alpha E_{h\cdot m}\left(\int_0^{\sigma_{X-F_n}} e^{-(\alpha+\gamma)t} f(X_t)dA_t\right) \uparrow \int_{F_n^{(0)}} h(x)f(x)\mu(dx)$, $\alpha \uparrow \infty$, $F_n^{(0)}$ being the fine interior. Note that the set $\bigcup_n F_n - \bigcup_n F_n^{(0)}$ is exceptional because of (5.1.7). By letting n tend to infinity, we get $\lim_{\alpha\to\infty} \alpha(h, U_A^{\alpha+\gamma} f) = \langle f\cdot\mu, h\rangle$, proving that A is related to μ by (5.1.9).

The uniqueness of A is then clear because $I_{F_n}\cdot A$ is related to $I_{F_n}\cdot\mu$ in the manner of Theorem 5.1.2 for each n. Q.E.D.

The proof of Theorem 5.1.3 is now completed. For $A \in \mathbf{A}_c^+$, the associated smooth measure (according to Theorem 5.1.3) is denoted by μ_A. The next lemma plays an important role in the subsequent sections.

Lemma 5.1.9. For any $A \in \mathbf{A}_c^+$, $\nu \in S_{00}$ and $t > 0$,

(5.1.14) $E_\nu(A_t) \leq (1+t)\|U_1\nu\|_\infty\cdot\mu_A(X)\quad (\leq \infty)$.

Proof. Assume first that $\mu = \mu_A$ is of finite energy integral: $\mu \in S_0$.

By setting $c_t(x) = E_x(A_t)$, $x \in X$, we claim that

(5.1.15) $\quad c_t \in \mathscr{F}$ and $\mathscr{E}(c_t, v) = \langle \mu, \tilde{v} - p_t\tilde{v}\rangle, \ \forall v \in \mathscr{F}$.

By Lemma 5.1.4, the potential $u_A^1(x)$ of A is a quasi-continuous version of the potential $U_1\mu$ of μ. Hence, $\langle \mu, c_t\rangle \leqq e^t\langle \mu, u_A^1\rangle = e^t \mathscr{E}_1(\mu) < \infty$. On the other hand, $(1/s)(c_t, c_t - p_s c_t) = (1/s)(c_t, c_s - p_t c_s) = (1/s)(c_t - p_t c_t, c_s)$ which is, by Lemma 5.1.4, equal to $(1/s)\int_0^s \langle \mu, p_u(c_t - p_t c_t)\rangle\, du = (1/s)\int_0^s \langle \mu, 2c_{t+u} - c_u - c_{2t+u}\rangle du$. Therefore, we get $\lim_{s \downarrow 0} (1/s)(c_t, c_t - p_s c_t) = \langle \mu, 2c_t - c_{2t}\rangle = \langle \mu, c_t - p_t c_t\rangle < \infty$, proving that $c_t \in \mathscr{F}$ and $\mathscr{E}(c_t, c_t) = \langle \mu, c_t - p_t c_t\rangle$. Similarly, we have the relation in (5.1.15).

For $\mu \in S_0$ and $\nu \in S_{00}$, (5.1.15) gives us $E_\nu(A_t) = \langle \nu, c_t\rangle = \mathscr{E}_1(U_1\nu, c_t) = \langle \mu, \widetilde{U_1\nu} - p_t(\widetilde{U_1\nu})\rangle + (c_t, U_1\nu) \leqq \|U_1\nu\|_\infty(\mu(X) + \int_X c_t(x)m(dx))$, which proves the inequality (5.1.14) by noting that $\sup_{t>0} (1/t)\int_X c_t(x)m(dx) = \lim_{t \downarrow 0} (1/t)\, E_m(A_t) = \mu(X)$.

When $\mu = \mu_A$ is a general smooth measure, we may consider the increasing sequence $\{F_n\}$ of closed sets of Theorem 3.2.3. Since $I_{F_n}\cdot A$ is related to $I_{F_n}\cdot\mu \in S_0$, we have $E_\nu((I_{F_n}\cdot A)_t) \leqq (1 + t)\, \|U_1\nu\|_\infty \cdot \mu(F_n)$. By letting n tend to infinity, we arrive at (5.1.14). Q.E.D.

Finally, we briefly mention the sweeping out of PCAF. Consider $\mu \in S_0$ and the associated $A \in \mathbf{A}_c^+$. $U_A^1 1$ is a quasi-continuous version of $U_1\mu$. Take any Borel set $D \subset X$. By virtue of Lemma 4.4.1, the function $H_1^D U_A^1 1$ is then a quasi-continuous version of the potential $U_1\mu_D$, μ_D being the 1-sweeping out of μ on D (see the end of §3.3.).

Since $\mu_D \in S_0$, we can consider a PCAF B associated with μ_D. B is called the 1-*sweeping out* of A on the set D. Thus,

(5.1.16) $\quad U_B^1 1(x) = H_1^D U_A^1 1(x)$ q.e.

B_t does not increase on the complementary set of D in the following sense:

(5.1.17) $\quad P_x(B_t = 0, \ \forall t < \sigma_D) = 1$ q.e. $x \in X$.

Indeed, we see from Theorem 4.2.3 that $H_1^D H_1^D f(x) = H_1^D f(x)$ q.e. $x \in X$ for any $f \in \mathscr{B}^+(X)$ (cf. [14; I, (11.9)]). Therefore, (5.1.16) implies $E_x(\int_0^{\sigma_D} e^{-t} dB_t) = U_B^1 1(x) - H_1^D U_B^1 1(x) = 0$ q.e. $x \in X$.

Example 5.1.1. Let $\mathbf{M} = (\Omega, \mathscr{M}, X_t, P_x)$ be the Brownian motion on R^n. The associated Dirichlet form $\mathscr{E}$ on $L^2(R^n)$ is given by $(1/2\mathbf{D}, H^1(R^n))$(see Example 4.3.1). The absolutely continuous measure

$$(5.1.18) \quad \mu(dx) = f(x)dx, \quad f \geqq 0,$$

is smooth with respect to the form $\mathscr{E}$ when f is locally integrable. In fact, μ is then a positive Radon measure charging no set of zero capacity since μ vanishes on any set of Lebesgue measure zero. The PCAF of **M** associated with μ by Theorem 5.1.3 is

$$(5.1.19) \quad A_t(\omega) = \int_0^t f(X_s(\omega))ds.$$

The same assertions hold even when μ diverges around a set of zero capacity. Suppose that $n \geqq 2$. Then, for example,

$$(5.1.20) \quad f(x) = |x|^\alpha, \quad x \in R^n,$$

gives a smooth measure for any real α. Indeed, $F_n = \{x \in R^n\colon\ 1/n \leqq |x| \leqq n\}$ satisfies not only (3.2.10), (3.2.11), but also (3.2.12) because the one point set $\{0\}$ is of zero Newtonian capacity (see Example 3.3.1). The associated PCAF A_t is still given by (5.1.19). A_t satisfies the finiteness condition

$$(5.1.21) \quad P_x\left(\int_0^t |X_s|^\alpha ds < \infty\right) = 1, \quad t > 0,$$

whenever $x \neq 0$, since $\{0\}$ is polar (see Example 4.3.1). Because of the law of the iterated logarithm, however, (5.1.21) is violated for $x = 0$ when $\alpha \leqq -2$. Thus A_t is not necessarily a PCAF in the ordinary sense. But it is always a PCAF in our sense because we can ignore the polar set $\{0\}$ as an exceptional set of A_t.

When $n = 1$, the Brownian motion on R^1 admits no non-empty exceptional set (Example 4.5.1). Hence, the present definition of AF reduces to the ordinary one. By virtue of the remark at the end of Example 3.3.3, we easily see that a measure μ on R^1 is smooth if and only if μ is a positive Radon measure. In particular, the δ-measure $\delta_{\{y\}}$ concentrated at one-point y admits a PCAF $2l(t,y,\omega)$ according to Theorem 5.1.3. $l(t,y)$ is called the *Brownian local time*. It is known that there exists a set $\Lambda \subset \Omega$ with $P_x(\Lambda) = 1$, $\forall x \in R^1$, such that for each $\omega \in \Lambda$, $l(t,y,\omega)$ is continuous in two variables t and y.[1)]

Notice that the formula of Lemma 5.1.4(iii) serves as a characterization of the correspondence of $\mu \in S$ and $A \in \mathbf{A}_c^+$. By integrating the corresponding formula for $\mu = \delta_{\{y\}}$ and $A_t = 2l(t,y)$, we get

$$(5.1.22) \quad \int_0^t I_E(X_s)ds = 2\int_E l(t,y)dy, \quad E \subset R^1.$$

Furthermore, we get

$$(5.1.23) \quad A_t(\omega) = 2\int_{R^1} l(t,y,\omega)\mu(dy)$$

1) Cf. N. Ikeda-S. Watanabe [45].

for the PCAF A associated with any positive Radon measure μ on R^1.

§5.2 Decomposition of Additive Functionals of Finite Energy

Let X, m and $\mathbf{M}$ be as in the preceding section. In the following three sections we deal with those AF's of the process $\mathbf{M}$ which are not necessarily non-negative.

For any AF A_t of $\mathbf{M}$, we set

$$(5.2.1) \quad e(A) = \lim_{t \downarrow 0} \frac{1}{2t} E_m(A_t^2)$$

whenever the limit exists. $e(A)$ is called the *energy* of A. First of all, we shall exhibit three important classes of AF's of finite energy.

(I) AF's generated by functions. Suppose that a function u on X possesses a version $\tilde{u}(\tilde{u} = u\ m\text{-a.e.})$ such that $\tilde{u}$ is finely continuous q.e. and finite q.e. Then

$$(5.2.2) \quad A_t^{[u]} = \tilde{u}(X_t) - \tilde{u}(X_0), \quad t > 0$$

defines an AF in our sense and, indeed, a unique equivalence class independent of the choice of the version $\tilde{u}$ by virtue of Lemma 4.2.5.

$A^{[u]}$ is well defined for any $u \in \mathscr{F}$ because we may take as $\tilde{u}$ a quasi-continuous version of u (Theorem 4.3.2). Moreover, $A_t^{[u]}$ is of finite energy and

$$(5.2.3) \quad e(A^{[u]}) = \mathscr{E}^{\text{res}}(u, u), \quad u \in \mathscr{F}$$

according to the formula (4.5.22). Here $\mathscr{E}^{\text{res}}$ is the resurrected form defined by (4.5.20). $\mathscr{E}^{\text{res}}$ equals $\mathscr{E}$ if and only if there is no killing inside X.

Suppose that the Dirichlet space $(\mathscr{F}, \mathscr{E})$ of the process $\mathbf{M}$ is transient in the sense of §1.5. Let $(\mathscr{F}_e, \mathscr{E})$ be the extended Dirichlet space. Then the relation (5.2.3) extendes to all $u \in \mathscr{F}_e$. Indeed, as we saw at the end of §3.1, each $u \in \mathscr{F}_e$ admits a quasi-continuous version $\tilde{u}$, by which the AF $A^{[u]}$ is still well defined. On the other hand, we can see in the same way as the proof of Lemma 1.5.4 that $\tilde{u}$ satisfies the relation (4.5.7). Hence, (5.2.3) together with (4.5.5) extends to $u \in \mathscr{F}_e$.

(II) Martingale AF's of finite energy. Consider the family

(5.2.4) $\mathscr{M} = \{M\colon M \text{ is an AF such that for each } t > 0\ E_x(M_t^2) < \infty$ and $E_x(M_t) = 0$ q.e. $x \in X\}$.

Since $E_m(M_t^2)$ is subadditive in t, $e(M)$ is well defined and

(5.2.5) $$e(M) = \sup_{t>0} \frac{1}{2t} E_m(M_t^2) \quad (\leq \infty)$$

for any $M \in \mathscr{M}$. We set

(5.2.6) $\mathring{\mathscr{M}} = \{M \in \mathscr{M}\colon e(M) < \infty\}$.

$M \in \mathscr{M}$ if and only if M is a square integrable ordinary perfect AF with mean zero of the Hunt process $\mathbf{M}|_{X-N}$, N being some properly exceptional set (depending on M in general). $(M_t, \mathscr{F}_t, P_x)_{t\geq 0}$ is then a martingale for each $x \in X - N$. Hence, we can utilize a method of P.A.Meyer [67; III, Théorème 3] to conclude that there exists for each $M \in \mathscr{M}$ a unique PCAF $\langle M \rangle \in \mathbf{A}_c^+$ such that for any $t > 0$

(5.2.7) $E_x(\langle M \rangle_t) = E_x(M_t^2)$ q.e. $x \in X$.[1)]

We call $M \in \mathscr{M}$ a *martingale AF(MAF)*. $\langle M \rangle$ is called its quadratic variation. Denote by $\mu_{\langle M \rangle}$ the smooth measure associated with $\langle M \rangle$ according to Theorem 5.1.3. $\mu_{\langle M \rangle}$ is called the *energy measure* of the MAF M. From (5.2.1), (5.2.7), and (5.1.9), we see that the energy of an MAF is just half of the total mass of its energy measure:

(5.2.8) $$e(M) = \frac{1}{2}\mu_{<M>}(X), \quad M \in \mathscr{M}.$$

(III) CAF's of zero energy. Consider the family

(5.2.9) $\mathscr{N}_c = \{N\colon N \text{ is a continuous AF, } e(N) = 0,\ E_x(|N_t|) < \infty$ q.e. for each $t > 0\}$.

The quadratic variation of $N \in \mathscr{N}_c$ vanishes in the following sense:

(5.2.10) $$\sum_{k=1}^{nT} (N_{(k+1)/n} - N_{k/n})^2 \to 0, \quad n \to \infty, \quad \text{in } L^1(P_m)$$

1) Meyer's construction gives an ordinary "non-perfect" PCAF $\langle M \rangle$ of $\mathbf{M}|_{X-N}$. Then $\langle M \rangle$ can be modified to be an element of $\mathbf{A}_c^+$ by "perfection" (see P.A. Meyer [68] and references therein).

because the expectation of the left hand side equals

$$\sum_{k=1}^{nT} E_m(E_{X_{R^l}}{}^n(N_{1/n}^2)) \leqq nT \cdot E_m(N_{1/n}^2) \to 0, \quad n \to \infty.$$

An example of $N \in \mathcal{N}_c$ is given by

(5.2.11) $N_t = \int_0^t f(X_s)ds$

for any Borel function $f \in L^2(X;m)$. Then $E_x(\int_0^t |f(X_s)| ds) \leqq e^t R_1 |f|(x)$ is finite q.e.[1] and N is a CAF. Furthermore, $E_m(N_t^2) = 2E_m(\int_0^t f(x_s)\int_s^t f(X_v)\, dvds) = 2\ E_m(\int_0^t \int_0^{t-s} f(X_s)p_v f(X_s)\, dvds) = 2\int_0^t \int_0^{t-s} (p_s 1, f \cdot p_v f)\, dvds$. Hence,

(5.2.12) $\frac{1}{2t} E_m(N_t^2) = \frac{1}{t}\int_0^t (p_{t-s}1, f \cdot S_s f)ds, \ S_s f(x) = \int_0^s p_v f(x) dv.$

Since the right hand side of (5.2.12) is dominated by $\frac{1}{t}\int_0^t sds \cdot (f, f)$, N of (5.2.11) is of zero energy, i.e., $N \in \mathcal{N}_c$.

We are particularly interested in the sum of the classes (II) and (III):

(5.2.13) $\mathscr{A} = \mathring{\mathscr{M}} \oplus \mathcal{N}_c$

namely, $\mathscr{A}$ consists of AF's A such that

(5.2.14) $A_t = M_t + N_t, \quad M \in \mathring{\mathscr{M}}, N \in \mathcal{N}_c.$

Clearly $\mathscr{A}$ is a linear space of AF's of finite energy. Moreover, the expression (5.2.14) of $A \in \mathscr{A}$ is unique because $\mathring{\mathscr{M}} \cap \mathcal{N}_c = \{0\}$ where 0 denotes the additive functional identically zero. In fact, if $A \in \mathring{\mathscr{M}}$ is of zero energy, then $\mu_{\langle A \rangle}$ vanishes by (5.2.8) and so does $\langle A \rangle$. Hence, $A = 0$ by (5.2.7).

We define the *mutual energy* of A, $B \in \mathscr{A}$ by

(5.2.15) $e(A, B) = \lim_{t \downarrow 0} \frac{1}{2t} E_m(A_t B_t).$

We know by Schwarz inequality that $e(A, B) = 0$ when either A or B is in $\mathcal{N}_c$. Therefore,

---

1) Since $\langle \nu, R_1 f^2 \rangle \leqq \|U_1\nu\|_\infty \cdot (f, f) < \infty$, $^\forall \nu \in S_{00}$, Theorem 3.3.2 implies that $(R_1|f|)^2 \leqq R_1 f^2$ is finite q.e.

(5.2.16) $e(A) = e(M)$ if $A = M + N$, $M \in \mathring{\mathscr{M}}$, $N \in \mathscr{N}_c$.

The main purpose of this section is to show that any AF $A^{[u]}$ ($u \in \mathscr{F}$) in the first class (I) actually belongs to $\mathscr{A}$, i.e., $A^{[u]}$ is expressed as a sum of elements of $\mathring{\mathscr{M}}$ and $\mathscr{N}_c$ uniquely. To this end we first prove that the space $\mathring{\mathscr{M}}$ is a real Hilbert space with inner product (5.2.15).

Theorem 5.2.1. $\mathring{\mathscr{M}}$ is a real Hilbert space with inner product e. More specifically, for any e-Cauchy sequence $M^{(n)} \in \mathring{\mathscr{M}}$, there exist a unique $M \in \mathring{\mathscr{M}}$ and a subsequence n_k such that $\lim_{n\to\infty} e(M^{(n)} - M) = 0$ and for q.e. $x \in X$, $P_x(\lim_{n_k\to\infty} M_t^{(n_k)} = M_t$ uniformly on any finite interval of $t) = 1$.

Proof. By virtue of Lemma 5.1.9 and identities (5.2.7), (5.2.8), we have

(5.2.17) $E_\nu(M_t^2) \leq 2(1 + t)\, \|U_1\nu\|_\infty e(M)$, $\nu \in S_{00}$.

$(M_t, P_\nu)_{t\geq 0}$ is then a square integrable martingale and Doob's inequality implies

(5.2.18) $P_\nu(\sup_{0\leq s\leq T} |M_s| > \alpha) \leq \frac{2}{\alpha^2}(1 + T)\|U_1\nu\|_\infty e(M)$.

Assume that $M^{(n)} \in \mathring{\mathscr{M}}$ constitutes an e-Cauchy sequence and select a subsequence $n_k \to \infty$ such that $e(M^{(n_k+1)} - M^{(n_k)}) < \frac{1}{2^{3k}}$.

By (5.2.18), $P_\nu\left(\sup_{0\leq s\leq T} |M_s^{(n_k+1)} - M_s^{(n_k)}| > \frac{1}{2^k}\right) \leq \frac{2(1+T)\,\|U_1\nu\|_\infty}{2^k}$.
Applying the Borel-Cantelli lemma to this inequality, we see

(5.2.19) $P_\nu(\Lambda_0) = 1$, $\nu \in S_{00}$

where $\Lambda_0 = \{\omega \in \Omega;\ M_s^{(n_k)}$ converges uniformly in s on each finite interval$\}$.

In view of Theorem 3.3.2, (5.2.19) implies $P_x(\Lambda_0) = 1$ q.e. $x \in X$. Denote by Γ_k a defining set of $M^{(n_k)}$ and put $\Lambda_1 = \bigcap_k \Gamma_k$. Furthermore, we set $M_s(\omega) = \lim_{n_k\to\infty} M_s^{(n_k)}(\omega)$, $s \geq 0$, for $\omega \in \Lambda_0 \cap \Lambda_1$. M_s is then an AF with defining set $\Lambda_0 \cap \Lambda_1$. Since $M_t^{(n)}$ is $L^2(P_\nu)$ convergent to M_t by virtue of (5.2.17), we see $E_\nu(M_t^2) < \infty$ and $E_\nu(M_t) = 0$ for any $\nu \in S_{00}$. It is easy to conclude from this with the help of Theorem 3.3.2 that $E_x(M_t^2) < \infty$ and $E_x(M_t) = 0$ q.e. $x \in X$, that is, $M \in \mathscr{M}$. For any $\varepsilon > 0$, choose N

such that $e(M^{(n)} - M^{(m)}) < \varepsilon, n, m > N$. We have $(1/2t)\, E_m((M_t^{(n)} - M_t)^2) \leq \varepsilon$ by Fatou's lemma and (5.2.5). Then $e(M^{(n)} - M) \leq \varepsilon$ by (5.2.5) again. Hence, $M \in \mathring{\mathscr{M}}$ and $M^{(n)}$ is e-convergent to M. Q.E.D.

Theorem 5.2.2. For any $u \in \mathscr{F}$, the AF $A^{[u]}$ can be expressed uniquely as

$$A^{[u]} = M^{[u]} + N^{[u]}, \quad M^{[u]} \in \mathring{\mathscr{M}}, \quad N^{[u]} \in \mathscr{N}_c. \tag{5.2.20}$$

Moreover, it holds that

$$e(M^{[u]}) = \mathscr{E}^{\text{res}}(u, u). \tag{5.2.21}$$

In the transient case, the above statements extend to any function u in the extended Dirichlet space $\mathscr{F}_e$.

Proof. We already know the uniqueness of the decomposition (5.2.20) and it suffices to show the existence of such $M^{[u]}$ and $N^{[u]}$.

We start with the special case where u is in the range of the resolvent: $u = R_1 f$, f being a Borel function in $L^2(X;m)$. In this case (5.2.20) is reduced to the usual semi-martingale decomposition, namely, we may set

$$\begin{cases} N_t^{[u]} = \int_0^t (u(X_s) - f(X_s))ds \\ M_t^{[u]} = u(X_t) - u(X_0) - N_t^{[u]}, \quad t \geq 0 \end{cases} \tag{5.2.22}$$

As we have seen, $N^{[u]}$ of (5.2.22) belongs to $\mathscr{N}_c$. Since u is an element of $\mathscr{F}$, $A_t^{[u]} = u(X_t) - u(X_0)$ is of finite energy and $M_t^{[u]}$ of (5.2.22) satisfies the relation (5.2.21) in view of (5.2.3). We easily see that $M^{[u]} \in \mathscr{M}$[1] and hence $M^{[u]} \in \mathring{\mathscr{M}}$.

Next, take any Borel function $u \in \mathscr{F}$ and define the approximating function u_n by

$$u_n = nR_{n+1}u = R_1 f_n \quad \text{with} \quad f_n = n(u - nR_{n+1}u). \tag{5.2.23}$$

By the uniqueness of the decomposition (5.2.20) for u_n's, we have then $M^{[u_n]} - M^{[u_m]} = M^{[u_n - u_m]}$ and

$$e(M^{[u_n]} - M^{[u_m]}) = \mathscr{E}^{\text{res}}(u_n - u_m, u_n - u_m). \tag{5.2.24}$$

---

1) Cf. footnote of pp. 132.

Since $u_n \in \mathscr{F}$ are $\mathscr{E}_1$-convergent to u and $\mathscr{E}_1(v, v)$ dominates $\mathscr{E}^{\text{res}}(v, v)$ for any $v \in \mathscr{F}$, we conclude that $M^{[u_n]}$ is an e-Cauchy sequence in the space $\mathring{\mathscr{M}}$.

By virtue of Theorem 5.2.1, the formula

$$(5.2.25)\quad \begin{cases} M^{[u]} = \lim_{n\to\infty} M^{[u_n]} \quad \text{in} \quad (\mathring{\mathscr{M}}, e) \\ N^{[u]} = A^{[u]} - M^{[u]}, u \in \mathscr{F} \end{cases}$$

makes sense and

$$(5.2.26)\quad M^{[u]} \in \mathring{\mathscr{M}}, e(M^{[u]}) = \mathscr{E}^{\text{res}}(u, u), \quad u \in \mathscr{F}.$$

It only remains to show that $N^{[u]}$ of (5.2.25) belongs to the space $\mathscr{N}_c$. Note that a subsequence n_k exists and

(5.2.27) $P_x(N_t^{[u_{n_k}]}$ converges to $N_t^{[u]}$ uniformly on any finite interval of $t) = 1$ q.e. $x \in X$,

because the same statements for $A^{[u_n]}$ and $M^{[u_n]}$ hold on account of Lemma 5.1.2 and Theorem 5.2.1 respectively. From this and (5.2.22), we know that $N^{[u]}$ is a CAF.

On the other hand, we have from (5.2.25)

$N_t^{[u]} = A_t^{[u-u_n]} - (M_t^{[u]} - M_t^{[u_n]}) + N_t^{[u_n]}$ and, consequently,

$$\overline{\lim_{t\downarrow 0}} \frac{1}{2t} E_m\left((N_t^{[u]})^2\right) \leq 3\, e(A^{[u-u_n]}) + 3\, e(M^{[u]} - M^{[u_n]}).$$

By virtue of (5.2.3), (5.2.24) and (5.2.25), the right hand side equals $6\,\mathscr{E}^{\text{res}}(u - u_n, u - u_n)$, which can be made arbitrarily small with large n. Therefore, $e(N^{[u]}) = 0$ and $N^{[u]} \in \mathscr{N}_c$.

When the process is transient, we can extend the decomposition (5.2.20) and the relation (5.2.21) to any $u \in \mathscr{F}_e$ in the same way as above by making use of an approximating sequence $u_n \in \mathscr{F}$ which is $\mathscr{E}$-convergent to u. Q.E.D.

Corollary 1. The following linearity and continuity hold:

(i) $M^{[au+bv]} = aM^{[u]} + bM^{[v]}, N^{[au+bv]} = aN^{[u]} + bN^{[v]}$, $a, b \in R^1, u, v \in \mathscr{F}$ ($u, v \in \mathscr{F}_e$ in transient case).

(ii) $u_n, u \in \mathscr{F}$ ($u_n, u \in \mathscr{F}_e$ in transient case), $\mathscr{E}(u_n - u, u_n - u) \to 0$, $n \to \infty \Rightarrow \exists$ a subsequence $\{u_{n_k}\}$, $\lim_{n_k\to\infty} M_t^{[u_{n_k}]} = M_t^{[u]}$, $\lim_{n_k\to\infty} N_t^{[u_{n_k}]} = N_t^{[u]}$ uniformly on each finite interval of t, P_x-a.s. for q.e. $x \in X$.

In fact, the relation (i) follows from the uniqueness of the decomposition (5.2.20). The equality (5.2.21) means that the linear transform

(5.2.28) $\Phi : u \rightsquigarrow M^{[u]}$

is continuous from the Dirichlet space $(\mathscr{F},\mathscr{E})$ (the extended Dirichlet space $(\mathscr{F}_e,\mathscr{E})$ in transient case) into the space $(\mathring{\mathscr{M}},e)$. Combining this with Theorem 5.2.1 and Lemma 5.1.2, we conclude the continuity statement (ii).

(5.2.21) also implies the following embedding statement telling us that the structure of the Dirichlet space may be studied entirely within the framework of the space $\mathring{\mathscr{M}}$ of martingales of finite energy.

Corollary 2. Suppose that the process **M** is transient and that there is no killing inside X. Then the transformation Φ of (5.2.28) is an isometry and the extended Dirichlet space $(\mathscr{F}_e, \mathscr{E})$ can be identified by Φ with a closed subspace of $(\mathring{\mathscr{M}}, e)$.

Having established decomposition (5.2.20), we are now in a position to study some basic properties of AF $M^{[u]}$ and $N^{[u]}$ in (5.2.20). In this section we give a theorem by which we can compute the energy measure of the martingale part $M^{[u]}$. Denote by $\mathscr{F}_b$ the family of all m-essentially bounded functions in $\mathscr{F}$.

Theorem 5.2.3. When $u \in \mathscr{F}_b$,

(5.2.29) $$\int_X \tilde{f}(x)\mu_{\langle M^{[u]}\rangle}(dx) = 2\mathscr{E}^{\text{res}}(u\cdot f, u) - \mathscr{E}^{\text{res}}(u^2, f), \quad f \in \mathscr{F}_b.$$

Proof. By (5.1.9) and (5.2.7),

(5.2.30) $$\int_X f(x)\mu_{\langle M^{[u]}\rangle}(dx) = \lim_{t\downarrow 0}\frac{1}{t} E_{f\cdot m}((M_t^{[u]})^2)$$

when f is a bounded γ-excessive function ($\gamma \geqq 0$) in $\mathscr{F}$. Since $M_t^{[u]} = \tilde{u}(X_t) - \tilde{u}(X_0) - N_t^{[u]}$ and $E_{f\cdot m}((N_t^{[u]})^2) \leqq \|f\|_\infty\, E_m((N_t^{[u]})^2)$, we can see that the right hand side of (5.2.30) equals

$$\lim_{t\downarrow 0}\frac{1}{t}\mathrm{E}_{f\cdot m}((\tilde{u}(X_t) - \tilde{u}(X_0))^2) = \lim_{t\downarrow 0}\left\{\frac{2}{t}(uf, u - p_t u) - \frac{1}{t}(u^2, f - p_t f)\right.$$
$$\left. - \frac{1}{t}(u^2 f, 1 - p_t 1)\right\} = 2\mathscr{E}(uf, u) - \mathscr{E}(u^2, f) - \int_X \tilde{u}^2 f\, dk,$$

which is equal to the right hand side of (5.2.29) in view of (4.5.20).

(5.2.29) is now valid for $f = \alpha R_\alpha g$, $g \in \mathscr{F}_b$. After letting $\alpha \to \infty$, we get (5.2.29) for general $f \in \mathscr{F}_b$. Q.E.D.

Example 5.2.1. Consider the case where X is an Euclidean domain D and $\mathbf{M}$ is an m-symmetric diffusion process on D whose Dirichlet form $\mathscr{E}$ on $L^2(D;m)$ possesses $C_0^\infty(D)$ as its core. We assume for simplicity that $\mathbf{M}$ admits no killing inside D. From Theorem 2.2.2 and Theorem 4.5.3, we know then that $\mathscr{E} = \mathscr{E}^{\text{res}}$ and $\mathscr{E}$ should have the expression

$$\mathscr{E}(u, v) = \sum_{i,j=1}^{n} \int_D \frac{\partial u(x)}{\partial x_i} \frac{\partial v(x)}{\partial x_j} \nu_{ij}(dx), \quad u, v \in C_0^\infty(D),$$

with some measures ν_{ij} satisfying (1.2.3). In §5.4, we give still another derivation of this formula using the stochastic calculus.

Now the right hand side of (5.2.29) equals

$$2 \sum_{i,j=1}^{n} \int_D \frac{\partial u(x)}{\partial x_i} \frac{\partial u(x)}{\partial x_j} f(x)\nu_{ij}(dx) \quad \text{for} \quad u, f \in C_0^\infty(D),$$

and thus (5.2.29) implies

$$(5.2.31) \quad \mu_{\langle M^{[u]}\rangle}(dx) = 2 \sum_{i,j=1}^{n} \frac{\partial u(x)}{\partial x_i} \frac{\partial u(x)}{\partial x_j} \nu_{ij}(dx), \quad u \in C_0^\infty(D).$$

Suppose that every ν_{ij} is absolutely continuous with respect to the Lebesgue measure. Take Borel density functions $a_{ij}(x)$ such that $\sum_{i,j=1}^{n} a_{ij}(x)\xi_i\xi_j \geq 0$, $\forall x \in D$, $\forall \xi \in R^n$. In view of Theorem 5.1.3, the identity (5.2.31) then implies

$$(5.2.32) \quad \langle M^{[u]}\rangle_t = 2\int_0^t \left(\sum_{i,j=1}^{n} \frac{\partial u}{\partial x_i} \frac{\partial u}{\partial x_j} a_{ij}\right)(X_s)ds, \quad u \in C_0^\infty(D).$$

Furthermore, if a_{ij} satisfies

$$(5.2.33) \quad \lambda_0|\xi|^2 \leq \sum_{i,j=1}^{n} a_{ij}(x)\xi_i\xi_j \leq \Lambda_0|\xi|^2, \quad \forall x \in D, \quad \forall \xi \in R^n,$$

for some positive constants λ_0 and Λ_0, and if $m(dx) = dx$, then (5.2.32) holds for any $u \in \mathscr{F}_b$ (namely, for any bounded function in $H_0^1(D)$) because so does (5.3.31).

§5.3 Continuous Additive Functionals of Zero Energy

We saw in §5.2 (III) that any AF in the class $\mathscr{N}_c$ is of zero quadratic variation in a certain sense. Nevertheless, $\mathscr{N}_c$ contains many CAF's which are not of bounded variation. In this section we concentrate our attention on the second term $N^{[u]} \in \mathscr{N}_c$ in the decomposition (5.2.20) of Theorem 5.2.2 and give some answers to the next basic questions:

(I) When is $N_t^{[u]}$ of bounded variation in t?
(II) Where is the support of $N_t^{[u]}$ located?
Let us start with the following characterization of $N^{[u]}$.

Theorem 5.3.1. The following conditions are equivalent for an AF A and for a function $u \in \mathscr{F}$ ($u \in \mathscr{F}_e$ in transient case):
(i) $A = N^{[u]}$,
(ii) $A \in \mathscr{N}_c$ and, for each $t > 0$,

(5.3.1) $E_x(A_t) = p_t\tilde{u}(x) - \tilde{u}(x)$, q.e. $x \in X$,

(iii) $A \in \mathscr{N}_c$, $\lim_{t\downarrow 0} E_x(A_t) = 0$, q.e. and

(5.3.2) $\lim_{t\downarrow 0} \frac{1}{t} E_{v\cdot m}(A_t) = -\mathscr{E}(u, v), \forall v \in \mathscr{F}.$

Proof. The implication (i) $\Rightarrow$ (ii) $\Rightarrow$ (iii) is clear. Assume that an AF A satisfies the condition (iii). Let us put $c_t(x) = E_x(A_t)$; then we see from $A \in \mathscr{N}_c$ and the relation

(5.3.3) $c_{t+s}(x) = c_t(x) + p_t c_s(x)$ q.e. for each $t, s > 0$,

that $c_t \in L^2(X; m)$, $t > 0$.

By (5.3.2), (5.3.3) and (1.5.5), we have that for any $v \in \mathscr{F}$, $t > 0$, and $T > 0$, $(v - p_T v, c_t) = \lim_{s\downarrow 0} 1/s\,(S_T v - p_s S_T v, c_t) = \lim_{s\downarrow 0} 1/s\,(S_T v - p_t S_T v, c_s) = -\mathscr{E}(S_T v - p_t S_T v, u) = -\mathscr{E}(S_T v, u - p_t u) = -(v - p_T v, u - p_t u)$. Therefore, $l_t = (v, c_t + u - p_t u)$ satisfies the equation $l_t = l_{t+T} - l_T$ of the linearity. Since $\lim_{t\downarrow 0} (1/t)\, l_t = 0$ by (5.3.2), $l_t = 0$ and, consequently,

(5.3.4) $c_t = p_t u - u$ m-a.e.

In particular, this means that $c_t \in \mathscr{F}$ ($c_t \in \mathscr{F}_e$ in transient case) and $\lim_{t\downarrow 0} c_t = 0$ in $\mathscr{F}$. Moreover, $c_t(x)$ is a q.e. limit of a quasi-continuous function $p_s c_{t-s}(x)$ as $s \downarrow 0$ in view of (5.3.3) and a condition in (iii). Since the convergence also takes place in the topology of $\mathscr{F}$, c_t is quasi-continuous and (5.3.4) is strengthened as (5.3.1) (see §3.1). The implication (iii) $\Rightarrow$ (ii) is proven.

Finally, we assume condition (ii) for A. In order to derive (i), we set $B_t = N_t^{[u]} - A_t = \tilde{u}(X_t) - \tilde{u}(X_0) - M_t^{[u]} - A_t$, $t > 0$, $q_t(x) = E_x(B_t^2)$ and claim that

(5.3.5) $q_t(x) = 0$ q.e. for each $t > 0$.

Clearly B_t is a CAF of zero energy and $E_x(B_t) = 0$ on account of the assumption (ii). But then $\int_X q_t(x)m(dx) = 0$ because the left hand side is subadditive in t. We thus have $q_t(x) = 0$ m-a.e., from which we can conclude $p_s q_t = 0$ q.e. for each $s > 0$. By using Fatou's lemma, $q_t(x) = E_x(\lim_{n\to\infty} (B_{t+1/n} - B_{1/n})^2) \leq \underline{\lim}_{n\to\infty} p_{1/n}q_t(x) = 0$ q.e., getting (5.3.5). Q.E.D.

An AF A is said to be *of bounded variation* if $A_t(\omega)$ is of bounded variation in t on each compact subinterval of $[0, \zeta(\omega))$ for every fixed ω in a defining set of A. A CAF A is of bounded variation if and only if A can be expressed as a difference of two PCAF's:

(5.3.6) $A_t(\omega) = A_t^{(1)}(\omega) - A_t^{(2)}(\omega),\ A^{(1)},\ A^{(2)} \in \mathbf{A}_c^+$.

If an AF A is continuous and *of bounded variation*, then its total variation $\{A\}_t (= \int_0^t |dA_s|)$ is a PCAF[1] and an expression (5.3.6) is provided by $A_t^{(1)} = \{A\}_t,\ A_t^{(2)} = \{A\}_t - A_t$ for instance.

For the sake of definiteness, we say that each measure ν in S_0 is *of finite 1-order energy integral.* In transient case, the same role is played by a measure ν *of finite* 0-*order energy integral* which is defined as a positive Radon measure satisfying

$$\int_X |v(x)|\nu(dx) \leq C\sqrt{\mathscr{E}(v, v)}, \quad v \in \mathscr{F} \cap C_0(X).$$

Let us say that a signed Radon measure ν on X is *of finite* 1-*order* (resp. 0-*order*) *energy integral* if so is the total variation $|\nu|$. This happens if and only if ν can be expressed as $\nu = \nu^{(1)} - \nu^{(2)}$, $\nu^{(1)}$ and $\nu^{(2)}$ being positive Radon measures of the same type. We can then define the α-potential of ν by $U_\alpha \nu = U_\alpha \nu^{(1)} - U_\alpha \nu^{(2)}$ (resp. 0-potential of ν by $U\nu = U\nu^{(1)} - U\nu^{(2)}$) which does not depend on the choice of $\nu^{(1)}$ and $\nu^{(2)}$. Moreover, the quasi-continuous version $\tilde{v}$ of any $v \in \mathscr{F}$ (resp. $v \in \mathscr{F}_e$) is integrable with respect to the total variation $|\nu|$ and

(5.3.7) $\mathscr{E}_\alpha(U_\alpha \nu, v) = \langle \nu, \tilde{v} \rangle \quad v \in \mathscr{F}$,

(5.3.7)′ $\mathscr{E}(U\nu, v) = \langle \nu, \tilde{v} \rangle \quad v \in \mathscr{F}_e$.

Given a signed Radon measure ν of finite 1-order (resp. 0-order in

1) The measurability of $\{A\}$ follows from Darboux's theorem (Cf, [76; §14]).

transient case) energy integral, we express it as $\nu = \nu^{(1)} - \nu^{(2)}$ in the above way and set

$$(5.3.8) \quad A = A^{(1)} - A^{(2)},$$

where $A^{(1)}$ and $A^{(2)}$ are PCAF corresponding to $\nu^{(1)}$ and $\nu^{(2)}$ respectively according to § 5.1. A of (5.3.8) is a CAF of bounded variation and does not depend on the choice of $\nu^{(1)}$ and $\nu^{(2)}$.

Lemma 5.3.1. For any signed Radon measure ν as above,

$$(5.3.9) \quad N_t^{[U_1\nu]} = \int_0^t \widetilde{U_1\nu}(X_s)\, ds - A_t, \quad t \geq 0,$$

and in the transient case,

$$(5.3.10) \quad N^{[U\nu]} = -A,$$

where A is the CAF of bounded variation associated with ν by (5.3.8).

Proof. In view of the linearity (Corollary 1 to Theorem 5.2.2), we may assume that ν is non-negative. Take $\nu \in S_0$ and a quasi-continuous version u of $U_1\nu$. By virtue of (5.2.25) and (5.2.27), we have then $N_t^{[u]} = \lim_{n\to\infty} \int_0^t (nR_{n+1}u(X_s) - f_n(X_s))\, ds$ with $f_n = n(u - nR_{n+1}u)$, the convergence being uniform in t on each finite interval P_x-a.s. for q.e. $x \in X$, by taking a subsequence if necessary. Hence the identity (5.3.9) follows from Lemma 5.1.2 and the method of the construction of $A \in \mathbf{A}_c^+$ carried out in the proof of Theorem 5.1.1.

In order to prove (5.3.10), observe that any positive Radon measure ν of finite 0-order energy integral is also an element of S_0 and

$$(5.3.11) \quad U\nu = U_1\nu + R(\widetilde{U_1\nu}),$$

where R denotes the 0-order resolvent of the process **M**. Let $f = \widetilde{U_1\nu}$; then $R_\alpha f (\in \mathscr{F})$ converges as $\alpha \downarrow 0$ to Rf in $(\mathscr{F}_e, \mathscr{E})$. Therefore, by the continuity (Corollary 1 to Theorem 5.2.2) and by an obvious modification of Lemma 5.1.2, we see

$$N_t^{[Rf]} = \lim_{\alpha\downarrow 0} N_t^{[R\alpha f]} = \lim_{\alpha\downarrow 0} \int_0^t (\alpha R_\alpha f(X_s) - f(X_s))ds = -\int_0^t f(X_s)ds.$$

(5.3.10) follows from this and (5.3.11). Q.E.D.

Before formulating an answer to the question (I) raised in the beginning of this section, we remind the readers of Theorem 3.2.3 and Lemma 5.1.6 (resp. their obvious modifications in transient case): if a non-negative Borel measure ν is smooth, then there exists an increasing sequence $\{F_k\}$ of closed set satisfying

$$(5.3.12)\quad P_x(\lim_{k\to\infty} \sigma_{X-F_k} < \zeta) = 0 \quad \text{q.e.} \quad x \in X,$$

$$(5.3.13)\quad I_{F_k}\cdot\nu \in S_0 (\text{resp. } I_{F_k}\cdot\nu \text{ is of finite 0-order energy integral}) \text{ for each } k.$$

Let us call such a sequence $\{F_k\}$ a *nest associated with* $\nu \in S$.

Given two smooth measures $\nu^{(1)}$ and $\nu^{(2)}$, we can always choose a common nest $\{F_k\}$ associated with $\nu^{(1)}$ and $\nu^{(2)}$. We then let

$$(5.3.14)\quad \nu_k = I_{F_k}\cdot\nu^{(1)} - I_{F_k}\cdot\nu^{(2)}, \quad k = 1, 2, \cdots,$$

and call ν_k the restriction to F_k of the difference $\nu^{(1)} - \nu^{(2)}$. Each ν_k is a signed Radon measure of finite 1-order energy integral (resp. of finite 0-order energy integral in the transient case). If we let

$$(5.3.15)\quad \mathscr{F}_k = \{u \in \mathscr{F} : \tilde{u} = 0 \text{ q.e. on } X - F_k\},$$

then by Theorem 4.4.1 and (5.3.7),

$$(5.3.16)\quad \mathscr{E}_\alpha(U_\alpha\nu_k - H_{\alpha,k}(\widetilde{U_\alpha\nu_k}), v) = \langle \nu_k, \tilde{v}\rangle, \quad \forall v \in \mathscr{F}_k,$$

where $H_{\alpha,k}f(x) = E_x(e^{-\alpha\sigma_k} f(X_{\sigma_k}))$, $\sigma_k = \sigma_{X-F_k}$.

The same statement for $\alpha = 0$ holds in the transient case. Keeping these in mind we proceed to the proof of the next theorem.

Theorem 5.3.2. The following two conditions are equivalent to each other for $u \in \mathscr{F}$ ($u \in \mathscr{F}_e$ in the transient case):

(i) $N^{[u]}$ is a CAF of bounded variation.

(ii) there exist smooth measures $\nu^{(1)}$ and $\nu^{(2)}$ such that

$$(5.3.17)\quad \mathscr{E}(u, v) = \langle \nu_k, \tilde{v}\rangle, \quad \forall v \in \mathscr{F}_k,$$

for every k. Here ν_k is the restriction to F_k of the difference $\nu^{(1)} - \nu^{(2)}$, $\{F_k\}$ being a common nest associated with $\nu^{(1)}$ and $\nu^{(2)}$. $\mathscr{F}_k$ is the space defined by (5.3.15).

Proof. We only give the proof for $u \in \mathscr{F}$. Suppose $N^{[u]} \in \mathscr{N}_c$ is of bounded variation; then $N^{[u]} = -A^{(1)} + A^{(2)}$ for some $A^{(1)}$ and $A^{(2)}$ $\in \mathbf{A}_c^+$. Denote by $\nu^{(1)}$ and $\nu^{(2)}$ the smooth measures associated with $A^{(1)}$ and $A^{(2)}$ respectively. Then the relation $\lim_{t \downarrow 0} (1/t)\ E_{v \cdot m}(N_t^{[u]}) = -\langle \nu_k, v \rangle$ holds for a function $v = h - H_{1,k}h\ (\in \mathscr{F}_k)$, h being any 1-excessive function in $\mathscr{F}$ in accordance with Theorem 5.1.3. Hence, the relation (5.3.17) is valid for such type of functions in view of Theorem 5.3.1. For any $v \in \mathscr{F}_k$, we can choose a sequence $\{v_n\}$ of the above type which is $\mathscr{E}_1$-convergent to v. Since ν_k is of finite 1-energy integral, v_n is $L^1(X; |\nu_k|)$-convergent to $\tilde{v}$ as well. Thus, (5.3.17) is established for any $v \in \mathscr{F}_k$ for each k.

Conversely, assume that the relation (ii) holds. Let $A^{(1)}$ and $A^{(2)}$ be the PCAF associated with the smooth measures $\nu^{(1)}$ and $\nu^{(2)}$ respectively. We set $A = -A^{(1)} + A^{(2)}$ and claim that

$$(5.3.18) \quad N^{[u]} = A.$$

Fix k and rewrite (5.3.17) as

$$(5.3.19) \quad \mathscr{E}_1(u, v) = \langle \mu_k, \tilde{v} \rangle \quad \forall v \in \mathscr{F}_k,$$

where $\mu_k = u \cdot m + \nu_k$. Since the left hand side of (5.3.19) equals $\mathscr{E}_1(u - H_{1,k}\tilde{u}, v)$ for any $v \in \mathscr{F}_k$, (5.3.16) and (5.3.19) imply that

$$(5.3.20) \quad u - H_{1,k}\tilde{u} = U_1\mu_k - H_{1,k}(\widetilde{U_1\mu_k}).$$

Now the CAF $I_{F_k} \cdot A = -I_{F_k} \cdot A^{(1)} + I_{F_k} \cdot A^{(2)}$ is associated with the measure $-\nu_k$. Hence, $B_t = \int_0^t u(X_s)ds - (I_{F_k} \cdot A)_t$ is associated with μ_k and, by Lemma 5.3.1, we arrive at the equality

$$(5.3.21) \quad N_t^{[U_1\mu_k]} = \int_0^t \{\widetilde{U_1\mu_k}(X_s) - u(X_s)\}\, ds + (I_{F_k} \cdot A)_t, \quad t > 0.$$

On the other hand, $H_{1,k}(\widetilde{U_1\mu_k})$ is also a 1-potential of a measure $\pi_k\mu_k$ (a difference of swept-out measures in S_0). Denote by $B^{(k)}$ the difference of PCAF's associated with $\pi_k\mu_k$ (we may call $B^{(k)}$ the sweeping out of B on the set $X - F_k$; See the last part of § 5.1). Then, once again using Lemma 5.3.1, we have

$$(5.3.22) \quad N_t^{[H_{1,k}(\widetilde{U_1\mu_k})]} = \int_0^t H_{1,k}(\widetilde{U_1\mu_k})(X_s)ds - B_t^{(k)}, \quad t > 0.$$

(5.3.20), (5.3.21), (5.3.22), and Corollary 1 to Theorem 5.2.2 imply

$$N_t^{[u-H_{1,k}\tilde{u}]} = -\int_0^t H_{1,k}\tilde{u}(X_s)ds + (I_{F_k}\cdot A)_t + B_t^{(k)}, \quad t > 0.$$

Since $B_t^{(k)} = 0$, $\forall t < \sigma_k$, by virtue of (5.1.17), we conclude

$$(5.3.23) \quad N_t^{[u-H_{1,k}\tilde{u}]} = -\int_0^t H_{1,k}\tilde{u}(X_s)ds + A_t, \quad t < \sigma_k.$$

It is easy to see that $H_{1,k}\tilde{u} \in \mathscr{F}$ constitutes an $\mathscr{E}_1$-Cauchy sequence. Since we may take as $\tilde{u}$ a quasi-continuous version of u in the restricted sense (see § 3.1), the property (5.3.12) of the nest $\{F_k\}$, Theorem 4.3.2, and the quasi-left continuity of the process **M** imply $\lim_{k\to\infty} H_{1,k}\tilde{u}(x) = E_x(e^{-\zeta}\tilde{u}(X_\zeta)) = 0$ q.e. $x \in X$, from which we get $\lim_{k\to\infty} \mathscr{E}_1(H_{1,k}\tilde{u}, H_{1,k}\tilde{u}) = 0$. We can now derive the desired equality (5.3.18) from (5.3.23) by letting k tend to infinity and by using (5.3.21), Corollary to Theorem 5.2.2, and Lemma 5.1.2. Q.E.D.

Corollary. The following conditions are equivalent to each other for $u \in \mathscr{F}$ (resp. $u \in \mathscr{F}_e$ in the transient case).

(i) u is a difference of two 1-excessive functions in $\mathscr{F}$ (resp. 0-excessive functions in $\mathscr{F}_e$).

(ii) $u = U_1\nu$ (resp. $u = U\nu$) with a signed Radon measure ν of finite 1-order (resp. 0-order) energy integral.

(iii) $N^{[u]}$ is a CAF of bounded variation and the associated measure $\nu^{(1)} - \nu^{(2)}$ in Theorem 5.3.2 has the additional property that $\lim_{k\to\infty} |\nu_k|$ is a positive Radon measure of finite 1-order (resp. 0-order) energy integral.

In fact, relation (i) $\Leftrightarrow$ (ii) and (ii) $\Rightarrow$ (iii) are implied in Theorem 3.2.1 and Lemma 5.2.1 respectively. (iii) $\Rightarrow$ (ii) is trivial.

Denote by $\mathscr{F}^*$ the dual space of the Dirichlet space $\mathscr{F}$ with metric $\mathscr{E}_1$. Each element $u \in \mathscr{F}$ defines a unique $T^{[u]} \in \mathscr{F}^*$ by

$$(5.3.24) \quad \mathscr{E}(u, v) = \langle T^{[u]}, v\rangle, \quad v \in \mathscr{F}.$$

Sometimes it is convenient to express the relation (5.3.2) as

$$(5.3.25) \quad \lim_{t\downarrow 0} \frac{1}{t} E_{\nu\cdot m}(A_t) = -\langle T^{[u]}, v\rangle, \quad v \in \mathscr{F}.$$

In the transient case, the dual space $\mathscr{F}_e^*$ of the extended Dirichlet space

$(\mathscr{F}_e, \mathscr{E})$ plays a more specific role in characterizing the CAF's $N^{[u]}$:

Theorem 5.3.3. Suppose the process **M** is transient. Then the family $\{N^{[u]}, u \in \mathscr{F}_e\}$ of CAF's stands in one-to-one correspondence with the dual space $\mathscr{F}_e^*$ of the extended Dirichlet space, the correspondence being characterized by the conditions that $A \in \mathscr{N}_c$, $\lim_{t \downarrow 0} E_x(A_t) = 0$ q.e. and

$$\lim_{t \downarrow 0} \frac{1}{t} E_{v \cdot m}(A_t) = -\langle T, v \rangle, \; v \in \mathscr{F}_e.$$

Moreover, A is of bounded variation if and only if $T \in \mathscr{F}_e^*$ is expressible as $T = \nu^{(1)} - \nu^{(2)}$, $\nu^{(1)}, \nu^{(2)} \in S$, in the sense of (5.3.17).

In fact, the relation (5.3.24) defines a one-to-one correspondence between $\mathscr{F}_e^*$ and $\mathscr{F}_e$ in this case. Hence, Theorem 5.3.3 follows from the preceding two theorems.

In order to formulate an answer to the question (ii) raised in the beginning of this section, recall the definition of the α-spectrum $\sigma_\alpha(u)$ of $u \in \mathscr{F}$ given in §3.3. In particular, a set $\sigma(u)$ is called the (0-) spectrum of $u \in \mathscr{F}$ if $\sigma(u)$ is the complement of the largest open set G such that $\mathscr{E}(u, v)$ vanishes for any $v \in \mathscr{F} \cap C_0(X)$ with Supp$[v] \subset G$.

Theorem 5.3.4. For any $u \in \mathscr{F}$ ($u \in \mathscr{F}_e$ in transient case), the AF $N_t^{[u]}$ vanishes on the complement of the spectrum $F = \sigma(u)$ of u in the following sense:

$$\text{(5.3.26)} \quad P_x(N_t^{[u]} = 0, \forall t < \sigma_F) = 1 \quad \text{q.e.} \quad x \in X.$$

Proof. First of all, consider a measure $\mu \in S_0$ and an AF $A \in \mathbf{A}_c^+$ which are related to each other by Theorem 5.1.3. Since $I_E \cdot \mu$ and $I_E \cdot A$ are related to each other for any Borel set E, we can say that the support of μ is just the complement of the largest open set G on which A vanishes in the sense that

$$\text{(5.3.27)} \quad P_x((I_G \cdot A)_t = 0, \; \forall t > 0) = 1, \text{ q.e. } x \in X.$$

Take $u \in \mathscr{F}$ and let $F = \sigma(u)$. Then $\sigma_1(u - G_1 u) \subset F$. By invoking the theorem of spectral synthesis (Theorm 3.3.4), we can find a sequence $\{\mu_n\}$ of signed Radon measures of finite (1-) energy integrals such that Supp $[\mu_n] \subset F$ and that the potentials $u_n = U_1 \mu_n$ are $\mathscr{E}_1$-convergent to $u - G_1 u$. By subtracting a subsequence if necessary, we then conclude from Corollary 1 to Theorem 5.2.2 that

$$\text{(5.3.28)} \quad \lim_{n \to \infty} N_t^{[u_n]} = N_t^{[u]} - N_t^{[G_1 u]}$$

uniformly on each finite interval of t P_x- a.s. for q.e. $x \in X$.

Denote by $A_t^{(n)}$ the CAF of bounded variation associated with μ_n in the manner of (5.3.8). Lemma 5.3.1 then implies $N_t^{[u_n]} = \int_0^t \tilde{u}_n(x_s)\, ds - A_t^{(n)}$, while Lemma 5.1.2 implies that $\int_0^t \tilde{u}_n(X_s)ds$ converges to $\int_0^t \tilde{u}(X_s)\, ds - \int_0^t \widetilde{G_1 u}(X_s)ds = -N_t^{[G_1 u]}$ in the same sense as above. Therefore, we get from (5.3.28) that

$$(5.3.29)\quad \lim_{n\to\infty} A_t^{(n)} = -N_t^{[u]},$$

the convergence being in the same sense as (5.3.28).

Since $A^{(n)}$ vanishes on the complement of F in the sense of (5.3.27), we are led from (5.3.29) to (5.3.26). Q.E.D.

The support Supp[A] of an AF A is defined by

$$(5.3.30)\quad \text{Supp}[A] = \{x \in X : P_x(R = 0) = 1\},$$

where $R(\omega) = \inf\{t > 0 : A_t(\omega) \neq 0\}$. Theorem 5.3.4 means

$$(5.3.31)\quad \text{Supp}\ [N^{[u]}] \subset \sigma(u),\ \forall u \in \mathscr{F}.$$

Example 5.3.1. Just as in Example 5.1.1, we consider the Brownian motion **M** on R^n. Let us examine the properties of CAF $N_t^{[u]}$ for $u \in H^1(R^n)$.

If the distribution derivative Δu belongs to $L^2(R^n)$, then

$$(5.3.32)\quad N_t^{[u]} = \frac{1}{2}\int_0^t \Delta u(X_s)ds,$$

which follows from Lemma 5.3.1 by the help of the expression $u = G_1(u - 1/2\Delta u)$ and Example 5.1.1.

In general, the following expression is valid:

$$(5.3.33)\quad N_t^{[u]} = \lim_{\varepsilon_n \downarrow 0} \frac{1}{2}\int_0^t \Delta u_{\varepsilon_n}(X_s)ds,\quad u \in H^1(R^n),$$

where $u_\varepsilon = \rho_\varepsilon * u$ is the convolution with the mollifier ρ_ε (see Problem 1.2.1), and $\{\varepsilon_n\}$ is a sequence decreasing to zero and depending only on the function u. The convergence in (5.3.33) is uniform on each finite interval of t P_x-a.s. for q.e. $x \in R^n$. Indeed, u_ε being convergent to u in the space $H^1(R^n)$ as $\varepsilon \downarrow 0$, Corollary 1 to Theorem 5.2.2 and identity (5.3.32) for $u = u_\varepsilon$ lead us to (5.3.33).

In view of Example 5.2.1, $M^{[u]}$ for $u \in H^1(R^n)$ is an MAF with the quadratic variation being given by

$$(5.3.34)\quad \langle M^{[u]}\rangle_t = \sum_{i=1}^{n} \int_0^t \left(\frac{\partial u(X_s)}{\partial x_i}\right)^2 ds.$$

In the next section we express $M^{[u]}$ by means of a stochastic integral. (5.3.32), (5.3.33), and (5.3.34) tell us that our decomposition $\tilde{u}(X_t) - \tilde{u}(X_0) = M_t^{[u]} + N_t^{[u]}$, $u \in H^1(R^n)$, is an extension of Ito's formula.

For any $u \in H^1(R^n)$ ($u \in \mathscr{F}_e$ when $n \geqq 3$), the AF $N^{[u]}$ vanishes (in the sense of Theorem 5.3.4) on the complement of the support of the distribution derivative Δu on account of Theorem 5.3.4 and (3.3.20). There are many $u \in H^1(R^n)$, however, for which $N_t^{[u]}$ are not of bounded variation as is shown in the discussion of the one-dimensional case below.

Suppose that $n \geqq 3$ and consider the extended Dirichlet space $\mathscr{F}_e$ given by (1.5.13). By virtue of Theorem 5.3.3, the family $\{N^{[u]}, u \in \mathscr{F}_e\}$ of AF's and the family $\{1/2\Delta u, u \in \mathscr{F}_e\}$ of distributions are in one-to-one correspondence. The latter family exhausts the dual space $\mathscr{F}_e^*$ described explicitly as follows: $\mathscr{F}_e^* = \{T$: tempered distribution, $\hat{T} \in L^2(|\xi|^{-2}d\xi)\}$.[1]

$N^{[u]}$ is of bounded variation if and only if $1/2\Delta u$ is represented by a difference of smooth measures in the sense of (5.3.17). This condition is certainly satisfied if $1/2\Delta u$ is a signed Radon measure of finite energy integral. But the converse is not necessarily true.

For instance, consider a signed measure $f(x)dx$ on R^3 with

$$(5.3.35)\quad f(x) = \begin{cases} |x|^\alpha \sin \dfrac{1}{|x|} & |x| < 1 \\ 0 & |x| \geqq 1. \end{cases}$$

$f(x)dx$ can always be expressed as a difference $f^+(x)dx - f^-(x)dx$ of smooth measures and

$$(5.3.36)\quad A_t = \int_0^t f(X_s)ds$$

defines a CAF of bounded variation of the 3-dimensional Brownian motion with a possible exceptional set being the origin 0 (see Example 5.1.1). The 0-order resolvent is now given by the Newtonian kernel $g(x,y) = \dfrac{1}{4\pi}\dfrac{1}{|x-y|}$ and the limit

$$(5.3.37)\quad u(x) = \lim_{\varepsilon\downarrow 0} \int_{|y|>\varepsilon} g(x,y)f(y)dy$$

1) Cf. Example 1.5.1. and J. Deny [25].

converges for every $x \in R^3$ whenever $\alpha > -3$. The Dirichlet integral of u equals

$$(5.3.38)\quad \lim_{\varepsilon \downarrow 0} \int_{|x|>\varepsilon, |y|>\varepsilon} g(x, y) f(x) f(y)\, dx dy.$$

When $-3 < \alpha$, (5.3.38) is convergent; consequently, the u defined in (5.3.37) belongs to $\mathscr{F}_e$. Moreover, $-N^{[u]}$ is given by (5.3.36) due to Corollary 1 of Theorem 5.2.2. When $-3 < \alpha \leqq -5/2$ however, (5.3.38) is not absolutely convergent, which means that $-1/2\Delta u = f dx$ can not be expressed as a difference of positive Radon measures of finite energy integrals. Nor can u be expressed as a difference of excessive functions belonging to $\mathscr{F}_e$ (Corollary to Theorem 5.3.2).

In the case of one-dimensional Brownian motion, a non-empty exceptional set is absent and the smooth measure reduces to the positive Radon measure (Example 5.1.1). This considerably simplifies the situation.

Thus $N^{[u]}$ for $u \in H^1(R^1)$ is of bounded variation if and only if the distribution $1/2u''$ is a signed Radon measure or equivalently u' is of bounded variation on each finite interval. When this is the case, $N_t^{[u]}$ is expressed as the local time integral:

$$(5.3.39)\quad N_t^{[u]} = \int_{R^1} l(t, y) du'(y), \quad P_x\text{-a.s.}, \quad x \in R^1.$$

In fact, from (5.1.22) we see that

$$\int_0^t u_\varepsilon''(X_s) ds = 2\int_{R^1} l(t, y) u_\varepsilon''(y) dy = 2\int_{R^1} l_\varepsilon(t, y) du'(y)$$

which combined with (5.3.33) lead us to (5.3.39). Here $l_\varepsilon(t, \cdot) = \rho_\varepsilon * l(t, \cdot)$.

As an example consider a function $g \in C_0(R^1)$ satisfying $\int_{R^1} g(x)dx = 0$ and set $u(x) = \int_{-\infty}^x g(y)dy$. Then $N^{[u]}$ is of bounded variation if and only if the function g is of bounded variation.

§5.4 Stochastic Calculus Related to the Dirichlet Form

A real-valued stochastic process Z_t is said to be a *semi-martingale* if Z_t is expressible as a sum of a (local) martingale and a process of bounded variation. In the last section we saw that our process $\tilde{u}(X_t) - \tilde{u}(X_0) = M_t^{[u]} + N_t^{[u]}$ for $u \in \mathscr{F}$ is not necessarily a semi-martingale. Accordingly the transformation rule of semi-martingales due to H. Kunita and S. Watanabe (a generalized Ito's formula) does not apply directly. Nevertheless, we can extend a due rule of transformation for the martingale part $M^{[u]}$. Our strategy is first to establish a transformation rule of the energy measures of $M^{[u]}$'s in a purely analytical manner and then to translate it in terms of a new notion of stochastic integrals.

Our assumptions for X, m and $\mathbf{M}$ are the same as in the preceding three sections. For simplicity the energy measure $\mu_{\langle M^{[u]}\rangle}$ of $M^{[u]} \in \mathring{\mathscr{M}}$ for $u \in \mathscr{F}$ is denoted by $\mu_{\langle u\rangle}$ and called the *energy measure* of u. When $u \in \mathscr{F}_b$ (an essentially bounded function in $\mathscr{F}$), (5.2.29) reads as follows:

$$\text{(5.4.1)} \quad \int_X \tilde{f}(x)\mu_{\langle u\rangle}(dx) = 2\mathscr{E}^{\text{res}}(u \cdot f, u) - \mathscr{E}^{\text{res}}(u^2, f), \quad f \in \mathscr{F}_b.$$

Actually the energy measure of $u \in \mathscr{F}_b$ can be introduced directly by this formula because the right hand side of (5.4.1), being equal to

$$\lim_{t\downarrow 0} \frac{1}{t} \int_{X\times X-d} (\tilde{u}(y) - \tilde{u}(x))^2 \tilde{f}(x) p_t(x, dy) m(dx)$$

as we saw in the proof of Theorem 5.2.3, defines a positive linear functional on $\mathscr{F} \cap C_0(X)$.

Since the total mass of $\mu_{\langle u\rangle}$ equals $2\mathscr{E}^{\text{res}}(u, u)$, we can define the bounded signed measure $\mu_{\langle u,v\rangle}$ for $u, v \in \mathscr{F}$ by

$$\mu_{\langle u,v\rangle} = \frac{1}{2}\{\mu_{\langle u+v\rangle} - \mu_{\langle u\rangle} - \mu_{\langle v\rangle}\}.$$

Then we have for $u, v, f \in \mathscr{F}_b$,

$$\text{(5.4.2)} \quad \int_X \tilde{f} d\mu_{\langle u,v\rangle} = \mathscr{E}^{\text{res}}(uf, v) + \mathscr{E}^{\text{res}}(vf, u) - \mathscr{E}^{\text{res}}(uv, f),$$

$$\text{(5.4.3)} \quad \int_X \tilde{f}\, d\mu_{\langle u,v\rangle} =$$

$$\lim_{t\downarrow 0} \frac{1}{t} \int_{X\times X-d} (\tilde{u}(y) - \tilde{u}(x))(\tilde{v}(y) - \tilde{v}(x)) \tilde{f}(x) p_t(x, dy) m(dx).$$

Lemma 5.4.1. For $u, v, f \in \mathscr{F} \cap C_0(X)$,

$$\text{(5.4.4)} \quad \int_X f\, d\mu_{\langle u^2,v\rangle} - 2\int_X fu\, d\mu_{\langle u,v\rangle}$$

$$= 2\int_{X\times X-d} (u(y) - u(x))^2 (v(y) - v(x)) f(x) J(dx, dy),$$

J being the jumping measure appearing in the Beurling-Deny first formula (Theorem 2.2.1 and Theorem 4.5.2).

Proof. By virtue of (5.4.3), the left hand side of (5.4.4) equals

$\lim_{t\downarrow 0} \frac{1}{t}\int_{X\times X-d} (u(y) - u(x))^2(v(y) - v(x))\, f(x)\, p_t(x, dy)\, m(dx) = \lim_{t\downarrow 0} \mathrm{I}(t)$ $- \lim_{t\downarrow 0} \mathrm{II}(t)$ where

$$\mathrm{I}(t) = \frac{1}{t}\int_{X\times X-d} (u(y) - u(x))^2(v(y) - v(x)) f(x)\phi(y)p_t(x, dy)m(dx),$$

$$\mathrm{II}(t) = \frac{1}{t}\int_{X\times X-d} u(x)^2 v(x) f(x)\, (1 - \phi(y))p_t(x, dy)m(dx),$$

ϕ being any non-negative function in $\mathscr{F} \cap C_0(X)$ such that $\phi = 1$ on a neighbourhood of Supp $[u]$ $\cup$ Supp $[v]$ $\cup$ Supp $[f]$.

It then holds that

$$\text{(5.4.5)}\quad \lim_{t\downarrow 0} \mathrm{I}(t) = 2\int_{X\times X-d} (u(y) - u(x))^2(v(y) - v(x)) f(x)\phi(y)J(dx, dy)$$

because the function $(v(y) - v(x))f(x)\phi(y)$ belongs to $C_\infty(X \times X - d)$ and the measure $(1/t)(u(y) - u(x))^2 p_t(x, dy)m(dx)$ is uniformly bounded by $2\mathscr{E}(u, u)$ (see the formula (4.5.7)) and vaguely converges to $2(u(y) - u(x))^2\, J(dx, dy)$ on $X\times X - d$ as $t \downarrow 0$.

On the other hand, we see from Lemma 4.5.2 and Theorem 4.5.2 that $\mathrm{II}(t) = \frac{1}{t}\,(u^2vf,\ \phi - p_t\phi) - \frac{1}{t}\int_X u(x)^2 v(x) f(x)\ (1 - p_t 1(x))\ m(dx)$ converges as $t \downarrow 0$ to $\mathscr{E}(u^2vf,\ \phi) - \int_X u(x)^2 v(x) f(x) k(dx) = \mathscr{E}^{\mathrm{res}}(u^2 v f,\ \phi)$ $= 2\int_{X\times X-d} u(x)^2 v(x) f(x)\ (1 - \phi(y))\ J(dx,\ dy)$, which, combined with (5.4.5), leads us to the desired identity (5.4.4). Q.E.D.

Lemma 5.4.2. Assume that the Dirichlet space $(\mathscr{F}, \mathscr{E})$ possesses the local property, or equivalently, that the jumping measure J vanishes identically. Then

$$d\mu_{\langle uv, w\rangle} = u\ d\mu_{\langle v, w\rangle} + v\ d\mu_{\langle u, w\rangle}, \qquad u, v, w \in \mathscr{F}_b.$$

Proof. It suffices to prove

$$\text{(5.4.6)}\quad \int_X f\, d\mu_{\langle u^2, v\rangle} = 2\int_X fu\ d\mu_{\langle u, v\rangle},\quad f \in \mathscr{F} \cap C_0(X),$$

for any $u, v \in \mathscr{F}_b$. This is true when $u, v \in \mathscr{F} \cap C_0(X)$ by virtue of the preceding lemma.

Fix $v \in \mathscr{F} \cap C_0(X)$ and take any $u \in \mathscr{F}_b$. Let $u_n \in \mathscr{F} \cap C_0(X)$ be $\mathscr{E}_1$-convergent to u. By virtue of Theorem 1.4.2, we may assume that u_n is uniformly bounded. Then by the same theorem and its proof, we can see that u_n^2 and $u_n^2 w$ are $\mathscr{E}^{\text{res}}$-weakly convergent to u^2 and $u^2 w$ respectively ($w \in \mathscr{F}_b$). Hence, the formula (5.4.2) implies

$$(5.4.7)\quad \lim_{n\to\infty} \int_X f\, d\mu_{\langle u_n^2, v\rangle} = \int_X f\, d\mu_{\langle u^2, v\rangle}, \quad f \in \mathscr{F} \cap C_0(X).$$

On the other hand, u_n converges to u q.e. by taking a subsequence if necessary (Theorem 3.1.4). Furthermore, the total variation of the signed measure $\mu_{\langle u_n - u, v\rangle}$ is dominated by

$$\sqrt{\mu_{\langle u_n - u\rangle}(X)}\,\sqrt{\mu_{\langle v\rangle}(X)} = \sqrt{2\mathscr{E}^{\text{res}}(u_n - u, u_n - u)}\,\sqrt{2\mathscr{E}^{\text{res}}(v, v)}$$

in view of the formula (5.4.3). Hence, $\left|\int_X f u_n\, d\mu_{\langle u_n, v\rangle} - \int_X f u\, d\mu_{\langle u, v\rangle}\right| \leq$ $\int_X |f u_n|\, |d\mu_{\langle u_n - u, v\rangle}| + \int_X |f|\, |u_n - u|\, |d\mu_{\langle u, v\rangle}| \to 0, \quad n \to \infty$, which, together with (5.4.7), implies (5.4.6) for $u \in \mathscr{F}_b$ and $v \in \mathscr{F} \cap C_0(X)$. In the same way, (5.4.6) extends to any $v \in \mathscr{F}_b$. Q.E.D.

We can now prove a transformation rule of energy measures. Let $\Phi(x) = \Phi(x_1, x_2, \ldots, x_m)$ be any continuously differentiable real function on $R^m (\Phi \in C^1(R^m))$ vanishing at the origin. For any $u_1, u_2, \cdots, u_m \in \mathscr{F}_b$, the composite function $\Phi(\mathbf{u}) = \Phi(u_1, u_2, \cdots, u_m)$ then belongs to $\mathscr{F}_b$ and

$$(5.4.8)\quad \sqrt{\mathscr{E}_\alpha(\Phi(\mathbf{u}), \Phi(\mathbf{u}))} \leq \sum_{i=1}^m \|\Phi_{x_i}\|_{L^\infty(V)} \sqrt{\mathscr{E}_\alpha(u_i, u_i)}, \quad \alpha \geq 0.$$

where V is an m-dimensional finite cube containing the range of $\mathbf{u}(x) = (u_1(x), \ldots, u_m(x))$, $x \in X$. If the derivatives Φ_{x_i} are uniformly bounded on R^m, then $\Phi(\mathbf{u}) = \Phi(u_1, u_2, \cdots, u_m)$ belongs to $\mathscr{F}$ for any $u_1, u_2, \cdots, u_m \in \mathscr{F}$ and

$$(5.4.9)\quad \sqrt{\mathscr{E}_\alpha(\Phi(\mathbf{u}), \Phi(\mathbf{u}))} \leq \sum_{i=1}^m \|\Phi_{x_i}\|_{L^\infty(R^m)} \sqrt{\mathscr{E}_\alpha(u_i, u_i)}, \quad \alpha \geq 0.$$

These can be shown in the same manner as the proof of Theorem 1.4.2, (ii).

Theorem 5.4.1. Assume that the Dirichlet space $(\mathscr{F}, \mathscr{E})$ possesses the local property. Then

$$(5.4.10)\quad d\mu_{\langle \Phi(\mathbf{u}), v\rangle} = \sum_{i=1}^m \Phi_{x_i}(\mathbf{u})\, d\mu_{\langle u_i, v\rangle}, \quad v \in \mathscr{F}_b,$$

for any $\Phi \in C^1(R^m)$ with $\Phi(\mathbf{o}) = 0$ and for any $u_1, u_2, \cdots, u_m \in \mathscr{F}_b$. This formula remains true for any $u_1, u_2, \cdots, u_m \in \mathscr{F}$ provided that Φ_{x_i} are uniformly bounded in addition.

Proof. Let $\mathscr{A}$ be the family of all Φ satisfying the first statement of the theorem. If $\Phi, \Psi \in \mathscr{A}$, then by Lemma 5.4.2, $d\mu_{\langle \Phi(u)\Psi(u), v\rangle} = \Phi(\mathbf{u})d\mu_{\langle \Psi(\mathbf{u}), v\rangle} + \Psi(\mathbf{u})\, d\mu_{\langle \Phi(\mathbf{u}), v\rangle} = \sum_{i=1}^{m} (\Phi(\mathbf{u})\Psi(\mathbf{u}))_{x_i}\, d\mu_{\langle u_i, v\rangle}$ for any $u_1, \cdots, u_m$, $v \in \mathscr{F}_b$, that is, the product $\Phi\Psi$ belongs to $\mathscr{A}$. Since $\mathscr{A}$ contains the coordinate functions, it contains all polynomials vanishing at the origin.

Take any $\Phi \in C^1(R^m)$ with $\Phi(\mathbf{o}) = 0$ and any $u_1, u_2, \cdots, u_m \in \mathscr{F}_b$. Let V be a finite cube containing the range of $\mathbf{u}(x) = (u_1(x), u_2(x), \cdots, u_m(x))$. Then there exists a sequence $\{\Phi^{(k)}(\mathbf{x})\}$ of polynomials vanishing at the origin such that $\Phi^{(k)} \to \Phi$, $\Phi^{(k)}_{x_i} \to \Phi_{x_i}$, $1 \leq i \leq m$, uniformly on V.[1)]

Due to the inequality (5.4.8), $\Phi^{(k)}(\mathbf{u})$ is then $\mathscr{E}_1$-convergent to $\Phi(\mathbf{u})$ as $k \to \infty$. Moreover, $\Phi^{(k)}(\mathbf{u}(x))$ is uniformly bounded and converges to $\Phi(\mathbf{u}(x))$, $x \in X$. Therefore, by letting k tend to infinity, we can get (5.4.10) in the same way as in the last part of the proof of the preceding lemma.

The same reasoning works in the proof of the second assertion of the theorem. Indeed, if we let $u_i^{(k)} = ((-k) \vee u_i) \wedge k$, $1 \leq i \leq m$, then (5.4.9) implies that $\Phi(\mathbf{u}^{(k)})$ is $\mathscr{E}$-convergent to $\Phi(\mathbf{u})$. Q.E.D.

Turning to the next task of introducing the notion of a stochastic integral, consider the space

$\dot{\mathbf{A}}_c = \{A - B : A, B \in \mathbf{A}_c^+, E_x(A_t) < \infty, E_x(B_t) < \infty, \forall t > 0, \text{q.e. } x \in X\}$.

For $M, L \in \mathscr{M}$, there exists a unique element $\langle M,L\rangle$ of $\dot{\mathbf{A}}_c$ such that

$$(5.4.11) \quad E_x(M_tL_t) = E_x(\langle M,L\rangle_t), \forall t > 0, \text{ q.e. } x \in X.$$

In fact, we may set $\langle M, L\rangle = (1/2)\{\langle M+L\rangle - \langle M\rangle - \langle L\rangle\}$. See the paragraph right after (5.2.6) for the uniqueness.

Let us consider the family

$\mathscr{M}_1 = \{M \in \mathscr{M} : \mu_{\langle M\rangle}(\in S) \text{ is a Radon measure}\}$.

$\mathscr{M}_1$ is a linear space containing the family $\mathring{\mathscr{M}}$ of MAF's of finite energy. There exists for $M, L \in \mathscr{M}_1$ a unique signed Radon measure $\mu_{\langle M,L\rangle}$ such that

$$(5.4.12) \quad \int_X f(x)\mu_{\langle M,L\rangle}(dx) = \lim_{t \downarrow 0} \frac{1}{t} E_m\left(\int_0^t f(X_s)d\langle M, L\rangle_s\right), \forall f \in C_0(X).$$

Indeed, it suffices to let $\mu_{\langle M,L\rangle} = (1/2)\{\mu_{\langle M+L\rangle} - \mu_{\langle M\rangle} - \mu_{\langle L\rangle}\}$.

---

1) Cf. R.Courant and D.Hilbert [18; Chap.II].

Lemma 5.4.3. If $M, L \in \mathcal{M}_1, f \in L^2(X; \mu_{\langle M\rangle})$ and $g \in L^2(X; \mu_{\langle L\rangle})$, then $f \cdot g$ is integrable with respect to the absolute variation $|\mu_{\langle M,L\rangle}|$ of $\mu_{\langle M,L\rangle}$ and

$$(5.4.13)\quad \left(\int_X |f \cdot g|\, |d\mu_{\langle M,L\rangle}|\right)^2 \leqq \int_X f^2 d\mu_{\langle M\rangle} \int_X g^2 d\mu_{\langle L\rangle}.$$

Proof. Making the expression $d\mu_{\langle M,L\rangle} = k_1 d\nu$, $d\mu_{\langle M\rangle} = k_2 d\nu$ and $d\mu_{\langle L\rangle} = k_3 d\nu$ with $\nu = \mu_{\langle M\rangle} + \mu_{\langle L\rangle} + |\mu_{\langle M,L\rangle}|$, we get from (5.4.11) and

(5.4.12), $d\mu_{\langle aM+bL\rangle} = (a^2 k_2 + 2abk_1 + b^2 k_3) d\nu$ for any $a, b \in R$.

Therefore, the set $B_0 = \bigcup_{a,b:\text{rational}} \{x \in X: a^2 k_2(x) + 2abk_1(x) + b^2 k_3(x) < 0\}$ is ν-negligible and we have for each $\alpha, \beta \in R$
$\alpha^2 f(x)^2 k_2(x) + 2\alpha\beta |f(x)g(x)k_1(x)| + \beta^2 g(x)^2 k_3(x) \geqq 0$ for every $x \in X - B_0$.
Integrating this with respect to ν,

$$\alpha^2 \int f^2 d\mu_{\langle M\rangle} + 2\alpha\beta \int |fg|\, |d\mu_{\langle M,L\rangle}| + \beta^2 \int g^2 d\mu_{\langle L\rangle} \geqq 0, \quad \text{proving (5.4.13).}$$

Q.E.D.

Theorem 5.4.2. Given $M \in \mathcal{M}_1$ and $f \in L^2(X; \mu_{\langle M\rangle})$, there exists a unique element $f \cdot M \in \mathring{\mathcal{M}}$ such that

$$(5.4.14)\quad e(f \cdot M, L) = \frac{1}{2} \int_X f(x) \mu_{\langle M,L\rangle}(dx), \ \forall L \in \mathring{\mathcal{M}}.$$

The mapping $f \rightsquigarrow f \cdot M$ is linear and continuous from $L^2(X; \mu_{\langle M\rangle})$ into the space $(\mathring{\mathcal{M}}, e)$.

Proof. By virtue of Lemma 5.4.3,
$$\left|\frac{1}{2} \int_X f d\mu_{\langle M,L\rangle}\right| \leqq \frac{1}{\sqrt{2}} \|f\|_{L^2(\mu_{\langle M\rangle})} \sqrt{e(L)}, \quad L \in \mathring{\mathcal{M}}.$$
Hence, Theorem 5.2.1 implies Theorem 5.4.2 together with the inequality $\sqrt{e(f \cdot M)} \leqq (1/\sqrt{2}) \|f\|_{L^2(\mu_{\langle M\rangle})}$. Q.E.D.

The $f \cdot M \in \mathring{\mathcal{M}}$ of Theorem 5.4.2 is called the *stochastic integral*. This terminology is legitimate by the following lemma.

Lemma 5.4.4. Let M, f and $f \cdot M$ be as in Theorem 5.4.2. Then

$$(5.4.15)\quad d\mu_{\langle f \cdot M, L\rangle} = f \cdot d\mu_{\langle M,L\rangle}, \quad L \in \mathring{\mathcal{M}}.$$

Moreover, the following approximation holds for $f \in C_0(X)$:

$$\text{(5.4.16)} \quad \lim_{|\Delta| \to 0} E_x(\{(f \cdot M)_t^{(\Delta)} - (f \cdot M)_t\}^2) = 0, \quad t > 0, \quad \text{q.e.} \quad x \in X$$

where

$$(f \cdot M)_t^{(\Delta)} = \sum_{i=1}^{n} f(X_{t_{i-1}})(M_{t_i} - M_{t_{i-1}}).$$

Δ denotes the partition $0 = t_0 < t_1 < \cdots < t_n = t$ and $|\Delta| = \max_{1 \leq i \leq n} (t_i - t_{i-1})$.

Proof. Let B be a common properly exceptional set for M and $\langle M \rangle$ such that the relation (5.2.7) holds for every $x \in X - B$. Applying the same argument as in [67; III, Théorèm 4] to the Hunt process $\mathbf{M}|_{X-B}$ and its AF's M and $\langle M \rangle$ in the ordinary sense, we can see for $f \in C_0(X)$ that there exists a unique martingale AF $\tilde{M}$ of $\mathbf{M}|_{X-B}$ in the ordinary sense[1)] satisfying for each $x \in X - B$,

$$\text{(5.4.17)} \quad \lim_{|\Delta| \to 0} E_x(\{(f \cdot M)_t^{(\Delta)} - \tilde{M}_t\}^2) = 0, \; E_x(\tilde{M}_t^2) = E_x\left(\int_0^t f(X_s)^2 d\langle M \rangle_s\right).$$

It follows from this that $e(\tilde{M}) = \lim_{t \downarrow 0} \frac{1}{2t} E_m(\tilde{M}_t^2) = \frac{1}{2} \int_X f^2 d\mu_{\langle M \rangle} < \infty$ and, consequently, $\tilde{M} \in \mathring{\mathcal{M}}$.

On the other hand, (5.4.11) and (5.4.17) imply

$$\text{(5.4.18)} \quad E_x(\langle \tilde{M}, L \rangle_t) = E_x\left(\int_0^t f(X_s) d\langle M, L \rangle_s\right)$$

for any $L \in \mathcal{M}$ and q.e. $x \in X$. When $L \in \mathring{\mathcal{M}}$, $\mu_{\langle \tilde{M}, L \rangle}$ is a bounded signed measure; hence, $\lim_{t \downarrow 0} \frac{1}{t} E_{h \cdot m}(\langle \tilde{M}, L \rangle_t) = \int_X h(x) \mu_{\langle M, L \rangle}(dx)$ for any bounded γ-excessive function h in view of the relation (5.1.9). Therefore, we have from (5.4.18)

$$\text{(5.4.19)} \quad \int_X h(x) \mu_{\langle \tilde{M}, L \rangle}(dx) = \int_X h(x) f(x) \mu_{\langle M, L \rangle}(dx), \quad L \in \mathring{\mathcal{M}}.$$

By setting $h = \alpha R_\alpha \tilde{h}$, $\tilde{h} \in C_0(X)$, and letting α tend to infinity, we conclude

1) $\tilde{M}$ can be made perfect by a method of "perfection" (see footnote of pp. 131). Thus, $\tilde{M} \in \mathcal{M}$.

that (5.4.19) holds for any $h \in C_0(X)$. Therefore, $d\mu_{\langle \tilde{M}, L\rangle} = f \cdot d\mu_{\langle M, L\rangle}$, $\forall L \in \mathring{\mathscr{M}}$. In particular, $e(\tilde{M}, L) = \frac{1}{2} \int_X f d\mu_{\langle M, L\rangle}$ which means $\tilde{M} = f \cdot M$.

We have proved Lemma 5.4.4 when $f \in C_0(X)$. On account of Lemma 5.4.3 and Theorem 5.4.2, the relation (5.4.19) for $\tilde{M} = f \cdot M$ readily extends to $f \in L^2(X; \mu_{\langle M\rangle})$. Q.E.D.

Corollary. (i) For $M \in \mathscr{M}_1$, $f \in L^2(X; \mu_{\langle M\rangle})$ and $g \in L^2(X; f^2\mu_{\langle M\rangle})$,

$$(5.4.20) \quad g \cdot (f \cdot M) = gf \cdot M.$$

(ii) For $M, L \in \mathscr{M}_1, f \in L^2(X; \mu_{\langle M\rangle})$ and $g \in L^2(X; \mu_{\langle L\rangle})$,

$$(5.4.21) \quad e(f \cdot M, g \cdot L) = \frac{1}{2} \int_X f(x) g(x) \mu_{\langle M, L\rangle}(dx).$$

By using Theorem 5.1.3, we can translate the formula (5.4.15) for $M \in \mathscr{M}_1$ and Borel $f \in L^2(X; \mu_{\langle M\rangle})$ into the following relationship for elements of $\dot{\mathbf{A}}_c$:

$$(5.4.22) \quad \langle f \cdot M, L\rangle = f \cdot \langle M, L\rangle, \ \forall L \in \mathring{\mathscr{M}}.$$

In this sense our stochastic integral $f \cdot M$ can be regarded as a variant of the stochastic integral due to M. Motoo and S. Watanabe. Note that

$$(5.4.23) \quad E_x\left(\int_0^t f(X_s)^2 d\langle M\rangle_s\right) < \infty \quad \text{q.e.} \quad x \in X$$

for M and f as above. In fact, we have by Lemma 5.1.9 that $E_\nu(\int_0^t f(X_s)^2 d\langle M\rangle_s) \leq (1+t) \| U_1\nu \|_\infty \cdot \int_X f(x)^2 \mu_{\langle M\rangle}(dx) < \infty$ for any $\nu \in S_{00}$. The Motoo-Watanabe integral has been formulated when the finiteness condition (5.4.23) is fulfilled for every $x \in X$ rather than for q.e. $x \in X$. See Example 5.4.1 below for a related discussion.

Let us extend the notion of the stochastic integral $f \cdot M$ to a wider class of AF's. We say that an AF M of the process $\mathbf{M}$ is *locally in* $\mathscr{M}_1$ ($M \in \mathscr{M}_{1,\text{loc}}$ in notation) if there exists a sequence $\{G_n\}$ of relatively compact open sets such that $\bar{G}_n \subset G_{n+1}$ and $G_n \uparrow X, n \to \infty$, and a sequence $\{M^{(n)}\}$ of MAF's in $\mathscr{M}_1$ such that $M_t = M_t^{(n)}$, $\forall t < \sigma_{X-G_n}$, P_x-a.s. for q.e. $x \in X$. The class $\mathring{\mathscr{M}}_{\text{loc}}$ of AF's is defined similarly. $\mathring{\mathscr{M}}_{\text{loc}}$ is a subclass of $\mathscr{M}_{1,\text{loc}}$.

If $M \in \mathscr{M}_{1,\text{loc}}$, its quadratic variation $\langle M\rangle \in \mathbf{A}_c^+$ is well difined by

(5.4.24) $\langle M\rangle_t = \langle M^{(n)}\rangle_t, \quad t < \sigma_{X-G_n}, \quad n = 1, 2, \cdots,$

by choosing an appropriate defining set and exceptional set of $\langle M\rangle$. $\langle M\rangle$ does not depend (up to the equivalence) on the special choice of $\{G_n\}$ and $\{M^{(n)}\}$ for M. For each $x \in X$ outside the exceptional set of M and for each n, the process $M^2_{t\wedge\sigma_n} - \langle M\rangle_{t\wedge\sigma_n}$ is a martingale with respect to P_x. Here $\sigma_n = \sigma_{X-G_n}$.

We can still define the energy measure of $M \in \mathscr{M}_{1,\mathrm{loc}}$ as the smooth measure $\mu_{\langle M\rangle}$ associated with the PCAF $\langle M\rangle$. Due to Lemma 5.1.5, we then have

(5.4.25) $\int_X f(x)\mu_{\langle M\rangle}(dx) = \int_X f(x)\mu_{\langle M^{(n)}\rangle}(dx), f \in C_0(X), \mathrm{Supp}[f] \subset G_{n-1}.$

In particular, $\mu_{\langle M\rangle}$ is a positive Radon measure and Lemma 5.4.3 extends to $M, L \in \mathscr{M}_{1,\mathrm{loc}}$.

Therefore, the formula (5.4.14) still provides us with the stochastic integral $f\cdot M \in \mathring{\mathscr{M}}$ for any $M \in \mathscr{M}_{1,\mathrm{loc}}$ and $f \in L^2(X; \mu_{\langle M\rangle})$. The relations (5.4.20), (5.4.21) and (5.4.22) extend to $\mathscr{M}_{1,\mathrm{loc}}$ as well. In particular, $(f\cdot M)_t$ together with $\langle f\cdot M\rangle_t$ vanishes until the first leaving time σ_{X-G} from a relatively compact open set G, P_x-a.s. for q.e. x, whenever f vanishes on the set G.

Keeping this observation in mind, we can finally define the stochastic integral $f\cdot M \in \mathring{\mathscr{M}}_{\mathrm{loc}}$ for $M \in \mathscr{M}_{1,\mathrm{loc}}$ and $f \in L^2_{\mathrm{loc}}(X; \mu_{\langle M\rangle})$ by the relation

(5.4.26) $g\cdot(f\cdot M) = (gf)\cdot M$

which holds for any bounded Borel function g with compact support. In fact, it suffices to let $(f\cdot M)_t = ((I_{G_n}\cdot f)\cdot M)_t$, $t < \sigma_{X-G_n}$, $\{G_n\}$ being any sequence of relatively compact open sets such that $\bar{G}_n \subset G_{n+1}$ and $G_n \uparrow X$, $n \to \infty$.

Here we mention a lemma which is useful in the subsequent developement.

Lemma 5.4.5. Let C_1 and $\mathscr{D}$ be a uniformly dense subfamily of $C_0(X)$ and an $\mathscr{E}_1$-dense subfamily of $\mathscr{F}$. Then the family $\{f\cdot M^{[u]}: f \in C_1, u \in \mathscr{D}\}$ of stochastic integrals is dense in $(\mathring{\mathscr{M}}, e)$.

Proof. Suppose that an MAF $M \in \mathring{\mathscr{M}}$ is e-orthogonal to the above family, namely, $\int_X f\, d\mu_{\langle M, M^{[u]}\rangle} = 0 \quad {}^\forall f \in C_1, {}^\forall u \in \mathscr{D}$. This identity extends to all $u \in \mathscr{F}$ in view of Lemma 5.4.3 and the equality $\mu_{\langle M^{[u]}\rangle}(X) = 2\mathscr{E}^{\text{res}}(u, u)$. Hence,

$$(5.4.27) \quad \langle M, M^{[u]}\rangle = 0 \quad \forall u \in \mathscr{F}.$$

In particular, (5.4.27) holds for $u = R_\alpha g$, $\alpha > 0$, $\forall g \in C_0(X)$. According to [67; III, Théorème 6], we then have $P_\nu(M = 0) = 1$, $\nu \in S_{00}$, from which follows $M = 0$. Q.E.D.

In the remainder of this section we assume that the process **M** is a diffusion: $P_x(X_t$ is continuous in $t < \zeta) = 1$, $x \in X$. In view of Theorem 4.5.1 the Dirichlet space $(\mathscr{F}, \mathscr{E})$ then possesses the local property and the jumping measure J vanishes identically. Accordingly,

$$(5.4.28) \quad \mathscr{E}^{\text{res}}(u, v) = \mathscr{E}^{(c)}(u, v), \; u, v \in \mathscr{F},$$

with the form $\mathscr{E}^{(c)}$ having the property (4.5.14) stronger than the local property.

For AF's $A^{(1)}, A^{(2)}$ and a stopping time σ, we write

$$A_t^{(1)} = A_t^{(2)} \quad \forall t < \sigma$$

when this is true P_x-a.s. for q.e. $x \in X$.

Lemma 5.4.6. If the difference of $u_1, u_2 \in \mathscr{F}$ is constant on a relatively compact open set G, then

$$(5.4.29) \quad M_t^{[u_1]} = M_t^{[u_2]} \quad \forall t < \sigma_{X-G}.$$

Proof. From (5.4.1) and (5.4.28) we see that for any $f \in \mathscr{F} \cap C_0(X)$, $\int_X f(x)\mu_{\langle u\rangle}(dx) = 0$ whenever $u \in \mathscr{F}_b$ and u is constant on a neighbourhood of Supp[f]. We can see that this statement holds for $u \in \mathscr{F}$ as well. To see this, observe the inequality $\left(\sqrt{\int_X |f|\, d\mu_{\langle u\rangle}} - \sqrt{\int_X |f|\, d\mu_{\langle v\rangle}}\right)^2 \leq \int_X |f|\, d\mu_{\langle u-v\rangle} \leq 2\|f\|_\infty \mathscr{E}^{\text{res}}(u - v, u - v)$, which follows from either (5.4.3) or (5.4.13). It then suffices to set $v = ((-n) \vee u) \wedge n$ and let n tend to infinity.

Therefore, $\mu_{\langle u_1 - u_2\rangle} = 0$ on G under the assumption of the present lemma. Then $\langle M^{[u_1]} - M^{[u_2]}\rangle_t = 0$ $\forall t < \sigma_{X-G}$, which means (5.4.29). Q.E.D.

We say that a function u is *locally in* $\mathscr{F}$ ($u \in \mathscr{F}_{\text{loc}}$ in notation) if for any relatively compact open set G there exists a function $w \in \mathscr{F}$ such that

$u = w$ m-a.e. on G. The space $\mathscr{F}_{b,loc}$ is defined in an analogous way. For $u \in \mathscr{F}_{loc}$, there exists a unique $M^{[u]} \in \mathring{\mathscr{M}}_{loc}$ such that

(5.4.30) $M_t^{[u]} = M_t^{[w]}, \ \forall t < \sigma_{X-G}$

for G and w as above. Indeed, Lemma 5.4.6 guarantees to set $M_t^{[u]} = M_t^{[u_n]}$ $\forall t < \sigma_{X-G_n}$, where $\{G_n\}$ is a sequence of relatively compact open sets such that $\bar{G}_n \subset G_{n+1}$, $G_n \uparrow X$, and $\{u_n\}$ are functions in $\mathscr{F}$ such that $u = u_n$ on G_n. In particular, we obtain from Lemma 5.4.6

(5.4.31) $u_1, u_2 \in \mathscr{F}_{loc}, u_1 - u_2 = \text{constant} \Rightarrow M^{[u_1]} = M^{[u_2]}$.

We can now present our transformation rule in the space $\mathring{\mathscr{M}}_{loc}$.

Theorem 5.4.3. Let $\mathbf{M}$ be an m-symmetric diffusion on X whose Dirichlet space $(\mathscr{F}, \mathscr{E})$ is regular.
(i) For any $\Phi \in C^1(R^m)$ and any $u_1, u_2, \cdots, u_m \in \mathscr{F}_{b,loc}$, the composite function $\Phi(\mathbf{u}) = \Phi(u_1, u_2, \ldots, u_m)$ is again locally in $\mathscr{F}_b$ and

(5.4.32) $M^{[\Phi(\mathbf{u})]} = \sum_{i=1}^{m} \Phi_{x_i}(\mathbf{u}) \cdot M^{[u_i]}$.

(ii) For any $\Phi \in C^1(R^m)$ with bounded dervatives and any $u_1, u_2, \cdots, u_m \in \mathscr{F}_{loc}$, the composite function $\Phi(\mathbf{u})$ is again locally in $\mathscr{F}$ and the formula (5.4.32) is satisfied.
(iii) Suppose $\Phi(\mathbf{u}) \in \mathscr{F}$ in the case (i) or (ii). Then

(5.4.33) $\mathscr{E}^{(c)}(\Phi(\mathbf{u}), \Phi(\mathbf{u})) = \frac{1}{2} \sum_{i,j=1}^{m} \int_X \Phi_{x_i}(\mathbf{u}) \Phi_{x_j}(\mathbf{u}) d\mu_{\langle u_i, u_j \rangle}$.

Proof. By virtue of Theorem 5.4.1, we have $\Phi(\mathbf{u}) \in \mathscr{F}_b$ and $\int_X fg d\mu_{\langle \Phi(\mathbf{u}), v \rangle} = \sum_{i=1}^{m} \int_X fg \Phi_{x_i}(\mathbf{u}) d\mu_{\langle u_i, v \rangle}$, $\forall f, g \in C_0(X)$, $\forall v \in \mathscr{F}_b$ for any $\Phi \in C^1(R^m)$ with $\Phi(\mathbf{o}) = 0$ and for any $u_1, u_2, \cdots, u_m \in \mathscr{F}_b$. This identity readily extends to any $\Phi \in C^1(R^m)$ and any $u_1, u_2, \ldots, u_m \in \mathscr{F}_{b,loc}$ with the help of (5.4.25) and (5.4.31). The isometry (5.4.21) then gives

$$e(f \cdot M^{[\Phi(\mathbf{u})]}, g \cdot M^{[v]}) = e\left(\sum_{i=1}^{m} f \Phi_{x_i}(\mathbf{u}) \cdot M^{[u_i]}, g M^{[v]}\right).$$

On account of Lemma 5.4.5, we have the following identity of the stochastic integrals in $\mathring{\mathscr{M}}$:

$$f\cdot M^{[\Phi(\mathbf{u})]} = \sum_{i=1}^{m} \{f\Phi_{x_i}(\mathbf{u})\}\cdot M^{[u_i]} \quad \forall f \in C_0(X),$$

which in turn means (5.4.32) by virtue of (5.4.26).

We have proved (i). (ii) can be proved in the same way. Under the condition of (iii), the energy of $M^{[\Phi(\mathbf{u})]}$ equals $\mathscr{E}^{(c)}(\Phi(\mathbf{u}),\Phi(\mathbf{u}))$ by (5.4.28). Thus the expression (5.4.33) results from the calculation of the energy of both sides of (5.4.32). Q.E.D.

Theorem 5.4.3 immediately leads us to the following presentation of the Beurling-Deny second formula:

Theorem 5.4.4. Let X be an Euclidean domain $D \subset R^n$ and $\mathbf{M}$ be an m-symmetric diffusion on D whose Dirichlet space $(\mathscr{F}, \mathscr{E})$ on $L^2(D; m)$ is regular. Suppose that the coordinate functions x_i are locally in $\mathscr{F}(1 \leq i \leq n)$. Then
(i) $C^1(D) \subset \mathscr{F}_{\text{loc}}$ and, for any $u \in C^1(D)$,

$$(5.4.34) \quad M^{[u]} = \sum_{i=1}^{n} u_{x_i}\cdot M^{(i)}, \quad M^{(i)} = M^{[x_i]}.$$

(ii) $C_0^1(D) \subset \mathscr{F}$ and, for any $u \in C_0^1(D)$,

$$(5.4.35) \quad \mathscr{E}^{(c)}(u, u) = \sum_{i,j=1}^{n} \int_D u_{x_i} u_{x_j} d\nu_{ij}, \quad \nu_{ij} = \frac{1}{2}\mu_{\langle x_i, x_j\rangle}.$$

Each measure ν_{ij} charges no set of zero capacity.

If we make the additional assumptions that $C_0^1(D)$ is $\mathscr{E}_1$-dense in $\mathscr{F}$ and that ν_{ij} vanishes identically for $i \neq j$, then the space $\mathring{\mathscr{M}}$ can be represented by the stochastic integrals:

$$(5.4.36) \quad \begin{cases} \mathring{\mathscr{M}} = \left\{\sum_{i=1}^{n} f_i\cdot M^{(i)} : f_i \in L^2(D; \nu_{ii}), 1 \leq i \leq n\right\} \\ e\left(\sum_{i=1}^{n} f_i M^{(i)}\right) = \sum_{i=1}^{n} \|f_i\|^2_{L^2(\nu_{ii})}. \end{cases}$$

Indeed, the right hand side of (5.4.36) is then a closed subsapce of $(\mathring{\mathscr{M}}, e)$, while its subclass $\{f\cdot M^{[u]} = \sum_{i=1}^{n} f\cdot u_{x_i}\cdot M^{(i)}, f \in C_0(D), u \in C_0^1(D)\}$ is dense in $(\mathring{\mathscr{M}}, e)$ according to Lemma 5.4.5.

Example 5.4.1 Consider the Brownian motion $\mathbf{M} = (X_t, P_x)$ on R^n (see

Examples 5.1.1 and 5.3.1) and set $B_t^{(i)} = X_t^{(i)} - X_0^{(i)}$, $X_t^{(i)}$ being the i-th coordinate of the sample path X_t $(1 \leqq i \leqq n)$. The coordinate functions x_i are in $H^1_{loc}(R^n)$, while

$$(5.4.37) \quad \begin{cases} M^{[x_i]} = B^{(i)} \in \mathscr{M}_1 \\ \langle B^{(i)}, B^{(j)} \rangle_t = \delta_{ij} \cdot t. \end{cases}$$

Hence $\mu_{\langle x_i, x_j \rangle} = \mu_{\langle B^{(i)}, B^{(j)} \rangle} = \delta_{ij} \cdot dx$; consequently, the stochastic integral $f \cdot B^{(i)}$ makes sense for every $f \in L^2(R^n)$. Furthermore,

$$(5.4.38) \quad \begin{cases} \mathring{\mathscr{M}} = \left\{ \sum_{i=1}^n f_i \cdot B^{(i)} : f_i \in L^2(R^n)\ 1 \leqq i \leqq n \right\} \\ e\left(\sum_{i=1}^n f_i \cdot B^{(i)}\right) = \frac{1}{2} \sum_{i=1}^n \| f_i \|^2_{L^2(R^n)} \end{cases}$$

in accordance with (5.4.36).

Thus the space $(\mathring{\mathscr{M}}, e)$ is identified with the space of vector fields $\mathbf{f} = (f_1, \cdots f_n)$, $f_i \in L^2(R^n)$. $\mathbf{f}$ represents $M^{[u]}$ for some $u \in H^1(R^n)$ if and only if $\mathbf{f}$ is a gradient field of an L^2-function and in this case

$$(5.4.39) \quad \mathbf{f} = \operatorname{grad} u.$$

When $n \geqq 3$, grad u is well defined for any u in the extended Dirichlet space $\mathscr{F}_e$ and $\{\operatorname{grad} u,\ u \in \mathscr{F}_e\}$ is a closed subspace of the above mentioned space of vector fields (see (1.5.18)). In this way the space $(\mathscr{F}_e, \frac{1}{2}\mathbf{D})$ is identified with a closed subspace of $(\mathring{\mathscr{M}}, e)$ (see Corollary 2 to Theorem 5.2.2).

We see from (5.4.23) that

$$(5.4.40) \quad E_x\left(\int_0^t f(X_s)^2 ds\right) < \infty \quad \text{q.e.} \quad x \in R^n,$$

for any $f \in L^2(R^n)$. For instance, the function

$$f(x) = f_0(x)/|x|^\alpha, \quad \alpha < \frac{n}{2}, \quad f_0 \in C_0(R^n),$$

belongs to $L^2(R^n)$. When $\alpha < 1$, (5.4.40) holds for every $x \in R^n$ and the Motoo-Watanabe stochastic integral $f \cdot B^{(i)}$ is well defined as an AF of $\mathbf{M}$ in the strict sense. When $\alpha \geqq 1$, however, $P_0(\int_0^t f(X_s)^2 ds = \infty) = 1$ (see Example 5.1.1) and neither the Motoo-Watanabe integral nor the Ito intgral is defined. The stochastic integral $f \cdot B^{(i)}$ in our sense is still defined by taking the polar set $\{0\}$ as its "essential' exceptional set.

Let us consider again an m-symmetric diffusion $\mathbf{M}$ on X whose Dirichlet space $(\mathscr{F}, \mathscr{E})$ is regular. Each $u \in \mathscr{F}_{loc}$ then admits a quasi-continuous version $\tilde{u}$. Furthermore, we may extend the decomposition (5.2.20) to $u \in \mathscr{F}_{\text{loc}}$ by setting

$$(5.4.41)\quad N_t^{[u]} = \tilde{u}(X_t) - \tilde{u}(X_0) - M_t^{[u]},\ t \geq 0,$$

and we may call $N^{[u]}$ a *CAF locally of zero energy*.

The notion of the α-spectrum still makes sense for $u \in \mathscr{F}_{\text{loc}}$ due to the local property of $\mathscr{E}$. An open set G is said to be an *α-regular set* of $u \in \mathscr{F}_{\text{loc}}$ if $\mathscr{E}_\alpha(w, v)$ vanishes for any $v \in \mathscr{F} \cap C_0(X)$ with $\text{Supp}[v] \subset G$ and for any $w \in \mathscr{F}$ such that $w = u$ on a neighbourhood of $\text{Supp}[v]$. The complementary set of the largest α-regular set of u is denoted by $\sigma_\alpha(u)$ and called the *α-spectrum* of $u \in \mathscr{F}_{\text{loc}}$ $(\alpha \geq 0)$.

It is easy to see that Theorem 5.3.4 concerning the support of $N^{[u]}$ remains true for any $u \in \mathscr{F}_{\text{loc}}$. In particular,

$$(5.4.42)\quad \tilde{u}(X_t) - \tilde{u}(X_0) = M_t^{[u]},\quad t \geq 0$$

when $\sigma_\alpha(u) = \phi$.

Example 5.4.2 Let $a_{ij}(x)$, $1 \leq i, j \leq n$, be locally integrable functions on an Euclidean domain $D \subset R^n$ such that $a_{ij} = a_{ji}$ and condition (5.2.33) is satisfied. We consider the case that $m(dx) = dx$. Let $\mathbf{M}$ be a symmetric diffusion on D whose Dirichlet space $(\mathscr{F}, \mathscr{E})$ is

$$(5.4.43)\quad \begin{cases} \mathscr{E}(u, v) = \sum_{i,j=1}^{n} \int_D \frac{\partial u}{\partial x_i} \frac{\partial v}{\partial x_j} a_{ij}(x)dx, & u,\ v \in C_0^\infty(D) \\ \mathscr{F} = H_0^1(D). \end{cases}$$

We have seen in Example 5.2.1 the following expression of the energy measure $\mu_{\langle u\rangle}$ of $u \in \mathscr{F}_b$:

$$d\mu_{\langle u\rangle} = 2 \sum_{i,j=1}^{n} \frac{\partial u}{\partial x_i} \frac{\partial u}{\partial x_j} a_{ij}dx.$$

This expression readily extends to any $u \in \mathscr{F}$ by the same reasoning as in the proof of Lemma 5.4.6. Therefore, the expression (5.2.32) of the quadratic variation of the MAF $M^{[u]}$ holds for any $u \in \mathscr{F}_{\text{loc}}$:

$$(5.4.44)\quad \langle M^{[u]}\rangle_t = 2\int_0^t \left(\sum_{i,j=1}^{n} \frac{\partial u}{\partial x_i} \frac{\partial u}{\partial x_j} a_{ij} \right)(X_s)ds.$$

Suppose $u \in \mathscr{F}_{\text{loc}}$ is harmonic in the sense that the right hand side of (5.4.43) vanishes for u and for every $v \in C_0^\infty(D)$, then the equality (5.4.42) is valid in virtue of Problem 3.3.4.

If the transition function of the diffusion **M** were absolutely continuous with respect to the underlying measure m, then any exceptional set is polar (§4.2). On account of the additivity and Markovity, we see that any harmonic $u \in \mathscr{F}_{\mathrm{loc}}$ satisfies for each $t \geqq s > 0$

$$(5.4.45) \quad \tilde{u}(X_t) - \tilde{u}(X_s) = M_t^{[u]} - M_s^{[u]}, \quad P_x\text{-a.s.}$$

for every $x \in D$ rather than for q.e. $x \in D$. An analogous statement is valid for the identity (5.4.44).

§5.5. Random Time Changes

In this section we assume that the process **M** is transient, i.e.,

$$Rf(x)\left(= \int_0^\infty p_t f(x)dt\right) < \infty \quad m\text{-a.e.}, \ \forall f \in \mathscr{B}^+ \cap L^1(X; m).$$

Thus, we start with an m-symmetric Hunt process **M** whose Dirichlet space $(\mathscr{F}, \mathscr{E})$ on $L^2(X; m)$ is not only regular but also transient. According to §1.5, the space $\mathscr{F}$ gives rise to the extended Dirichlet space $\mathscr{F}_e$ by means of the completion with respect to the metric $\mathscr{E}$. Moreover, each element of $\mathscr{F}_e$ admits a quasi-continuous version by virtue of regularity $(\mathscr{F}_e.3)'$ of §1.5 and the argument in §3.1.

Throughout this section every function in the space $\mathscr{F}_e$ is considered to be quasi-continuous already. In view of condition $(\mathscr{F}_e.2)$ of §1.5 and Lemma 3.1.4, we then see that $\mathscr{E}(u, u) = 0$ if and only if $u = 0$ q.e. Thus, two functions in $\mathscr{F}_e$ are identified if they coincide q.e.

Let us fix an arbitrarily positive Radon measure μ on X charging not set of zero capacity. Let A be the PCAF associated with μ by Theorem 5.1.3. We denote by Y and $\tilde{Y}$ the support of μ and the support of A respectively; i.e., Y is the smallest closed set outside which μ vanishes, while

$$(5.5.1) \quad \tilde{Y} = \{x \in X - N : P_x(R = 0) = 1\}.$$

Here N is a properly exceptional set of A and $R(\omega) = \inf\{t > 0: A_t(\omega) > 0\}$.

In general, it is hard to identify the set $\tilde{Y}$ with Y, but we can get the following relation by choosing a properly exceptional set N of A appropriately.

Lemma 5.5.1. $\tilde{Y}$ is a Borel set, $\tilde{Y} \subset Y$, and $\mu(Y - \tilde{Y}) = 0$.

Proof. Note that $\tilde{Y} = \{x \in X - N:\ \phi_A(x) = 1\}$ with $\phi_A(x) =$

$E_x(e^{-R})$. Since $\phi_A(x)$ is 1-excessive on $X - N$, we may assume by Lemma 4.2.6 that $\tilde{Y}$ is a Borel set by taking a larger properly exceptional set N of A if necessary.

It is known that $P_x((I_{X-N-\tilde{Y}} \cdot A)_t = 0) = 1$, $t > 0$, $x \in X - N$([14; Chap V, (3.9)]). In view of Theorem 5.1.3, we have then $\mu(X - \tilde{Y}) = \mu(X - N - \tilde{Y}) = 0$, and consequently,

(5.5.2) $\mu(Y - \tilde{Y}) = 0.$

On the other hand, we have

(5.5.3) $\mathrm{Cap}(\tilde{Y} - Y) = 0.$

Indeed, Theorem 5.1.3 implies that $A = I_Y \cdot A$ and, consequently, $P_x(A_{\sigma_Y} = 0) = 1$ q.e. Hence $P_x(\sigma_Y \leq R) = 1$ q.e. and $P_x(0 < R) = 1$ q.e. on an open set $X - Y$, which means (5.5.3).

We can now make the set $\tilde{Y} - Y$ empty by choosing a larger properly exceptional set N of A including $\tilde{Y} - Y$. Q.E.D.

Let us consider the following orthogonal decomposition of the Hilbert space $(\mathscr{F}_e, \mathscr{E})$:

(5.5.4) $\mathscr{F}_e = \mathscr{F}_{X-\tilde{Y}} \oplus \mathscr{H}^{\tilde{Y}}, \ \mathscr{F}_{X-\tilde{Y}} = \{u \in \mathscr{F}_e : u = 0 \text{ q.e. on } \tilde{Y}\}.$

We also define the space $\mathscr{H}^Y$ by replacing $\tilde{Y}$ in (5.5.4) with Y. We note that $\mathscr{H}^{\tilde{Y}}$ is identical with $\mathscr{H}^Y$ if and only if

(5.5.5) $\mathrm{Cap}(Y - \tilde{Y}) = 0.$

However we do not know whether (5.5.5) is true in general.

By virtue of the 0-order variant of Theorem 4.4.1,

(5.5.6) $u = \tilde{H}(\tilde{\gamma}u), \quad u \in \mathscr{H}^{\tilde{Y}},$

where $\tilde{\gamma}$ denotes the restriction to the set $\tilde{Y}$ and $\tilde{H}\phi(x) = E_x(\phi(X\sigma_{\tilde{Y}}))$. The hitting measure $H^{\tilde{Y}}(x, \cdot)$ is concentrated on the set $\tilde{Y}$ for q.e. $x \in X$ on account of Theorem 4.2.3. Our present aim is to identify the space $\mathscr{H}^{\tilde{Y}}$, by means of the operations $\tilde{\gamma}$ and $\tilde{H}$, with the extended Dirichlet space on Y associated with the time changed process of **M**.

Consider the time changed process $\mathbf{M}^{\tilde{Y}} = \{\Omega, \mathscr{M}, \mathscr{M}_{\tau_t}, X_{\tau_t}, P_x\}_{x \in \tilde{Y}}$ where

$$\tau_t(\omega) = \inf\{s > 0; A_s(\omega) > t\}. \tag{5.5.7}$$

It is known that $\mathbf{M}^{\tilde{Y}}$ is a normal, right continuous strong Markov process on $\tilde{Y}$ ([14; Chap V, (2.11)]).

Theorem 5.5.1. (i) The transition function of the time changed process $\mathbf{M}^{\tilde{Y}}$ is μ-symmetric and determines a strongly continuous transient semigroup $\{T_t^\mu, t > 0\}$ on $L^2(Y; \mu)$.
(ii) Let $(\mathscr{F}_e^\mu, \mathscr{E}^\mu)$ be the corresponding extended Dirichlet space with reference measure μ. Then the operation $\tilde{\gamma}$ determines a unitary equivalence between the Hilbert spaces $(\mathscr{F}_e^\mu, \mathscr{E}^\mu)$ and $(\mathscr{H}^{\tilde{Y}}, \mathscr{E})$. More specifically

$$\mathscr{F}_e^\mu = \tilde{\gamma}\mathscr{H}^{\tilde{Y}} \tag{5.5.8}$$

in the sense that $\mathscr{F}_e^\mu \supset \tilde{\gamma}\mathscr{H}^{\tilde{Y}}$ and, conversely, for each $\phi \in \mathscr{F}_e^\mu$, there exists $u \in \mathscr{H}^{\tilde{Y}}$ with $\tilde{\gamma}u = \phi$ μ-a.e. When ϕ and u are so related,

$$\mathscr{E}^\mu(\phi, \phi) = \mathscr{E}(u, u). \tag{5.5.9}$$

Proof. (i) According to the 0-order version of Theorem 3.2.3, there exists a sequence $\{F_n\}$ of increasing closed sets satisfying (3.2.11) and (3.2.12) such that $I_{F_n} \cdot \mu$ is of finite 0-order energy integral for each n. Hence, there exists a μ-integrable bounded function h on Y strictly positive μ-a.e. such that $\mathscr{F}_e \subset L^1(Y; h \cdot \mu)$ and

$$\int_Y |u(x)| h(x)\mu(dx) \leq \sqrt{\mathscr{E}(u, u)}, \; u \in \mathscr{F}_e. \tag{5.5.10}$$

Set $\mathbf{B}_1(\tilde{Y}) = \{\phi = \psi \cdot h: \psi \text{ is a bounded Borel function on } \tilde{Y}\}$. (5.5.10) implies that the 0-order potential $U(\phi \cdot \mu) \in \mathscr{F}_e$ of the signed Radon measure $\phi \cdot \mu$ is well-defined for every $\phi \subset \mathbf{B}_1(\tilde{Y})$. Furthermore, we can see from Lemma 5.1.4, Theorem 5.1.3, and the resolvent equation that

$$\widetilde{U(\phi \cdot \mu)}(x) = E_x\left(\int_0^\infty \phi(X_t) dA_t\right), \quad \text{q.e. } x \in X, \phi \in \mathbf{B}_1(\tilde{Y}). \tag{5.5.11}$$

Now we denote by $\{\tilde{R}_\alpha, \alpha > 0\}$ the resolvent of the time changed process $M^{\tilde{Y}}$, i.e.,

$$\tilde{R}_\alpha \phi(x) = E_x\left(\int_0^\infty e^{-\alpha t} \phi(X_{\tau_t}) dt\right) = E_x\left(\int_0^\infty e^{-\alpha A_t} \phi(X_t) dA_t\right), \quad x \in \tilde{Y}. \tag{5.5.12}$$

We have from (5.5.11) and (5.5.12) that $U(\phi\cdot\mu)(x) = \tilde{H}(\tilde{R}_0\phi)(x)$ q.e. $x \in X$, $\phi \in \mathbf{B}_1(\tilde{Y})$.[1] Therefore,

$$(5.5.13)\quad \begin{cases} \tilde{H}(\tilde{R}_0(\mathbf{B}_1(\tilde{Y})) \subset \mathscr{H}^{\tilde{Y}} \\ \mathscr{E}(\tilde{H}(\tilde{R}_0\phi), v) = (\phi, v)_\mu, \quad \phi \in \mathbf{B}_1(\tilde{Y}), \quad v \in \mathscr{H}^{\tilde{Y}}, \end{cases}$$

where $(\ ,\ )_\mu$ denotes the inner product on $L^2(Y; \mu)$. In particular, $\tilde{R}_0$ is μ-symmetric since $\tilde{R}_0\phi = \tilde{H}(\tilde{R}_0\phi)$ q.e. on $\tilde{Y}$.

(5.5.13) readily extends to any $\phi \in \mathbf{B}(\tilde{Y})$ such that $(\phi, \tilde{R}_0\phi)_\mu$ is finite, while the resolvent equation for $\{\tilde{R}_\alpha\}$ implies $(\tilde{R}_\alpha\phi, \tilde{R}_0\tilde{R}_\alpha\phi)_\mu \leqq \alpha^{-1}(\tilde{R}_\alpha\phi, \tilde{R}_0\phi)_\mu = \alpha^{-1}(\tilde{R}_0\tilde{R}_\alpha\phi, \phi)_\mu \leqq \alpha^{-2}(\tilde{R}_0\phi, \phi)_\mu < \infty$, $\phi \in \mathbf{B}_1^+(\tilde{Y})$. Hence, $\tilde{H}(\tilde{R}_\alpha\phi) = -\alpha\tilde{H}(\tilde{R}_0\tilde{R}_\alpha\phi) + \tilde{H}(\tilde{R}_0\phi)$ belongs to $\mathscr{H}^{\tilde{Y}}$ and $\mathscr{E}(\tilde{H}(\tilde{R}_\alpha\phi), v) + \alpha(\tilde{R}_\alpha\phi, v)_\mu = (\phi, v)_\mu$, $\phi \in \mathbf{B}_1(\tilde{Y})$, $v \in \mathscr{H}^{\tilde{Y}}$. Consequently, $\tilde{R}_\alpha$ is μ-symmetric.

On account of the right continuity of the sample path of the process $\mathbf{M}^{\tilde{Y}}$, it holds for $\phi \in C_0(Y)$ that $\lim_{\alpha\to\infty} \alpha\tilde{R}_\alpha\phi(x) = \phi(x)$ for every $x \in \tilde{Y}$ and, consequently, for μ-a.e. $x \in Y$. Therefore Lemma 1.4.3 applies and $\{\tilde{R}_\alpha, \alpha > 0\}$ gives a strongly continuous resolvent on $L^2(Y; \mu)$. The same statement holds for the transition function of $\mathbf{M}^{\tilde{Y}}$, which is transient because $\tilde{R}_0 h = U(h\cdot\mu)$ is finite q.e. on $\tilde{Y}$, and, consequently, μ-a.e. on Y.

(ii) Since $(R_0\varphi, \varphi) < \infty$, $\varphi \in \mathbf{B}_1(\tilde{Y})$, by (5.5.13), Theorem 1.5.3 gives

$$(5.5.14)\quad \begin{cases} \tilde{R}_0(B_1(\tilde{Y})) \subset \mathscr{F}^\mu_e \\ \mathscr{E}^\mu(\tilde{R}_0\phi, \psi) = (\phi, \psi)_\mu, \ \phi \in \mathbf{B}_1(\tilde{Y}), \ \psi \in \mathscr{F}^\mu_e. \end{cases}$$

(5.5.13) and (5.5.14) readily imply that the isometry (5.5.9) holds for $\phi = \tilde{R}_0\tilde{\phi}$ and $u = \tilde{H}(\tilde{R}_0\tilde{\phi})$ where $\tilde{\phi} \in \mathbf{B}_1(\tilde{Y})$. Since $\tilde{R}_0(B_1(\tilde{Y}))$ is dense in $(\mathscr{F}^\mu_e, \mathscr{E}^\mu)$ and any Cauchy sequence in $(\mathscr{F}^\mu_e, \mathscr{E}^\mu)$ (resp. in $(\mathscr{H}^{\tilde{Y}}, \mathscr{E})$) contains a subsequence which converges μ-a.e. (resp. q.e.), it only remains to show that $\tilde{H}(\tilde{R}_0(B_1(\tilde{Y}))$ is dense in $(\mathscr{H}^{\tilde{Y}}, \mathscr{E})$.

Suppose $v \in \mathscr{H}^{\tilde{Y}}$ is $\mathscr{E}$-orthogonal to $\tilde{H}(\tilde{R}_0(\mathbf{B}_1(\tilde{Y}))$. Then $v = 0$ μ-a.e. by virtue of (5.5.13). Set $f_n = |v| \wedge n$. Theorem 5.1.3 then implies that the AF $f_n\cdot A$ vanishes identically. Letting β tend to infinity in the equality $\beta E_x(\int_0^\infty e^{-\beta t} f_n(X_{\tau_t})dt) = 0$ q.e., and noting the quasi-continuity of f_n and Theorem 4.3.2, we arrive at $f_n = 0$ q.e. on $\tilde{Y}$. Hence, $v = 0$ q.e. on $\tilde{Y}$ and $v = \tilde{H}(\gamma v) = 0$ q.e. on X. Q.E.D.

The Dirichlet space on $L^2(Y; \mu)$ (resp. the extended Dirichlet space with reference measure μ) associated with the time changed semigroup $\{T^\mu_t, t > 0\}$ on $L^2(Y; \mu)$ appearing in the above theorem is called the

1) Cf. [14; Chap.V, (3.5)].

time changed Dirichlet space on $L^2(Y; \mu)$ (resp. the *time changed extended Dirchlet space* with reference measure μ). A positive Radon measure μ on X is said to be a *speed measure* if $\mathrm{Supp}[\mu] = X$ and μ charges no set of zero capacity.

Given a speed measure μ, we denote by $\tilde{X}$ the support of the associated PCAF. $X - \tilde{X}$ is always μ-negligible by Lemma 5.5.1. But if we assume the stronger condition that

$$\text{(5.5.15)} \quad \mathrm{Cap}(X - \tilde{X}) = 0,$$

then the space $\mathscr{H}^{\tilde{X}}$ is identified with $\mathscr{F}_e$ and we can get from Theorem 5.5.1 the following simple statement concerning the time changed Dirichlet space and the time changed process.

Theorem 5.5.2. Let μ be a speed measure satisfying condition (5.5.15).
(i) The time changed extended Dirichlet space $(\mathscr{F}^\mu_e, \mathscr{E}^\mu)$ with respect to μ is identical with the original extended Dirichlet space $(\mathscr{F}_e, \mathscr{E})$; i.e., $\mathscr{F}_e \subset \mathscr{F}^\mu_e$, and conversely, each $u \in \mathscr{F}^\mu_e$ admits a $\tilde{u} \in \mathscr{F}_e$ such that $u = \tilde{u}$ μ-a.e. When u and $\tilde{u}$ are so related,

$$\text{(5.5.16)} \quad \mathscr{E}^\mu(u, u) = \mathscr{E}(\tilde{u}, \tilde{u}).$$

(ii) The time changed process $\mathbf{M}^{\tilde{X}}$ can be made to be a standard process on $\tilde{X}$, by choosing the properly exceptional set of the associated PCAF appropriately.

In fact a choice of a lager properly exceptional set N including $X - \tilde{X}$ yields the quasi-left continuity of the time changed process up to the life time. Under the condition of Theorem 5.5.2, the following description of the corresponding time changed Dirichlet space $(\mathscr{F}^\mu, \mathscr{E}^\mu)$ on $L^2(X; \mu)$ holds:

$$\text{(5.5.17)} \quad \begin{cases} \mathscr{F}^\mu = \mathscr{F}_e \cap L^2(X; \mu) \\ \mathscr{E}^\mu = \mathscr{E}\,|_{\mathscr{F}_e \cap L^2(X;\, \mu)}. \end{cases}$$

Let μ be a positive Radon measure on X which is mutually absolutely continuous with respect to the original underlying measure m, i.e.,

$$\text{(5.5.18)} \quad d\mu = f \cdot dm$$

with a locally m-integrable, strictly positive (m-a.e.) Borel function f on X. Then μ is a speed measure satisfying condition (5.5.15). Indeed the function ϕ_A in the proof of Lemma 5.5.1 equals 1 μ-a.e., and, consequently, m-a.e. By Lemma 4.2.5, $\phi_A = 1$ q.e., which means (5.5.15).

Example 5.5.1. Let **M** be the Brownian motion on R^n for $n \geqq 3$. **M** is transient and the corresponding function space $\mathscr{F}_e$ (we only consider quasi-continuous functions) is the space of BLD functions (see Example 3.1.1).

Let f be a locally integrable, strictly positive (a.e.) Borel function on R^n. Then

$$(5.5.19)\quad \begin{cases} \mathscr{F}^f = \mathscr{F}_e \cap L^2(R^n; fdx) \\ \mathscr{E}^f(u, v) = \dfrac{1}{2}\mathbf{D}(u, v), \quad u, v \in \mathscr{F}^f \end{cases}$$

is the Dirichlet space on $L^2(R^n; fdx)$ associated with the time changed process of the Brownian motion **M** by means of the PCAF (5.1.19).

Example 5.5.2 (Symmetric Cauchy process on the boundary).

Let $\hat{\mathbf{M}}$ be the reflecting barrier Brownian motion on the 3-dimensional demi-space $\bar{D} = \{x = (x_1, x_2, x_3); x_3 \geqq 0\}$ (see Example 4.4.2). The associated regular Dirichlet space relative to $L^2(\bar{D})$ is the Sobolev space (1/2 **D**, $H^1(D)$).

$\hat{\mathbf{M}}$ is transient by (5.5.19). The associated extended Dirichlet space $\mathscr{F}_e$ is the completion of $H^1(D)$ with respect to the Dirichlet integral. The set of all quasi-continuous functions in $\mathscr{F}_e$ is denoted by $\mathscr{F}_e$ again. $\mathscr{F}_e$ then admits the following orthogonal decomposition:

$$(5.5.20)\quad \mathscr{F}_e = \mathscr{F}_e^{(0)} \oplus \mathscr{H}$$

where $\mathscr{F}_e^{(0)}$ is the space of all BLD functions on D of potential type (see Example 3.1.1) and $\mathscr{H}$ is simply the space of all harmonic functions on D with finite Dirichlet integrals. The element of $\mathscr{F}_e$ is simply called the *BLD function.* As the 0-order version of (4.4.15), we have

$$(5.5.21)\quad \mathscr{F}_e^{(0)} = \{u \in \mathscr{F}_e : u = 0 \quad \text{q.e.} \quad \text{on } \partial D\}.$$

Let us consider the 2-dimensional Lebesgue measure μ on the boundary ∂D, i.e., $\mu(dx) = dx_1 dx_2 \delta_{\{0\}}(dx_3)$. μ charges no set of zero capacity (with respect to the Dirichlet form (1/2**D**, $H^1(D)$) because $I_K \cdot d\mu$ is of finite 1-order energy integral for each compact set $K \subset \partial D$ in view of the proof of Lemma 2.2.1.

Recall that the reflecting barrier Brownian motion $\hat{\mathbf{M}}$ is obtained from the 3-dimensional Brownian motion $((X_t^{(1)}, X_t^{(2)}, X_t^{(3)}), P_x)$ by means of the reflection $\hat{X}_t = (X_t^{(1)}, X_t^{(2)}, |X_t^{(3)}|)$. Denote by $l(t)$ the local time at 0 of the one dimensional Brownian motion $X_t^{(3)}$ (see Example 5.1.1). Let A be the PCAF of the process $\hat{\mathbf{M}}$ corresponding to the measure μ by Theorem 5.1.3. Then

$$(5.5.22)\quad A_t(\omega) = 4l(t, \omega).$$

In fact, by making use of the equality

$$E_{(x_1,x_2,x_3)}(2l(t)) = \int_0^t \frac{1}{\sqrt{2\pi s}} e^{-(x_3)^2/2s}\, ds,$$

we can see that the formula (iii) of Lemma 5.1.4 holds for $4l(t)$, μ, and the transition function of $\hat{\mathbf{M}}$.

In particular, we have

(5.5.23) $\quad \mathrm{Supp}[\mu] = \mathrm{Supp}[A] = \partial D.$

Furthermore, the time changed process $\mathbf{M}^{\partial D} = \{X_{\tau_t}, P_x\}_{x \in \partial D}$ of $\mathbf{M}$ with respect to the AF A_t is a symmetric Cauchy process on ∂D. More specifically, $\mathbf{M}^{\partial D}$ is a spatially homogeneous Hunt process on ∂D (see Example 4.1.1) such that

(5.5.24) $\quad E_{(0,0,0)}((\exp(i\xi \cdot X_{\tau_t})) = \exp\left(-\frac{|\xi|}{2}t\right),\ \xi \in R^2$

Hence, the Dirichlet space $(\mathscr{F}^{\partial D}, \mathscr{E}^{\partial D})$ of $\mathbf{M}^{\partial D}$ on $L^2(\partial D) = L^2(\partial D; dx_1 dx_2)$ is given by

(5.5.25) $$\begin{cases} \mathscr{E}^{\partial D}(u, v) = \dfrac{1}{2\pi}\displaystyle\int_{R^3 \times R^3} (u(x) - u(y))(v(x) - v(y)) \frac{dx\, dy}{|x - y|^3} \\ \mathscr{F}^{\partial D} = \left\{u \in L^2(\partial D) : \displaystyle\int_{R^3 \times R^3} (u(x + y) - u(x))^2 \frac{1}{|y|^3}\, dx\, dy < \infty\right\}. \end{cases}$$

in virtue of (1.4.25).

In fact, the inverse function τ_t of $4l(t)$ obeys the one-sided stable law of exponent 1/2, i.e., $E_0(\exp(-\alpha\tau_t)) = \exp(-\sqrt{\alpha/2}\, t)$ (see Ito-McKean[49]). Consequently,

$$E_0(\exp(i\xi X_{\tau_t})) = \int_0^\infty E_0(e^{i\xi \cdot X_s})\, P_0(\tau_t \in ds) = \int_0^\infty e^{-|\xi|^2 s/2}\, P_0(\tau_t \in ds) = e^{-|\xi|t/2}.$$

(5.5.20), (5.5.21), and (5.5.23) imply that Theorem 5.5.1 applies with $\tilde{Y} = Y = \partial D$ and $\mathscr{H}^{\tilde{Y}} = \mathscr{H}$. We conclude that the space $\mathscr{H}$ of harmonic functions on D with finite Dirichlet integrals and the extended Dirichlet space $(\mathscr{F}_e^{\partial D}, \mathscr{E}^{\partial D})$ of the symmetric Cauchy process $\mathbf{M}^{\partial D}$ on ∂D are unitary equivalent by the correspondence

(5.5.26) $$\left(\mathscr{H}, \frac{1}{2}\mathbf{D}\right) \underset{H}{\overset{\gamma}{\rightleftarrows}} (\mathscr{F}_e^{\partial D}, \mathscr{E}^{\partial D}).$$

Here γ denotes the restriction to ∂D and $H\phi(x) = E_x(\phi(X\sigma_{\partial D}))$.

An analogous relation has been derived in (1.2.19) of Example 1.2.3, when D is the unit disk on R^2.

CHAPTER 6

Construction of Symmetric Markov Processes

It is shown in this chapter that, given a regular Dirichlet space, there exists a Hunt process whose Dirichlet space is the given one. Such a process is unique up to an equivalence in accordance with Theorem 4.3.6. Furthermore, the local property of the given regular Dirichlet space is equivalent to the sample continuity of the constructed Hunt process by virtue of Theorem 4.5.1.

As was explained at the end of §4.1, any transition function possessing the Feller property admits a Hunt process. In general, it is hopeless to construct a Feller transition function from the L^2-semigroup T_t associated with the given Dirichlet space. However, if the Dirichlet space is regular, then the potential theory of Chapter 3 provides us with quasi-continuous versions $\widetilde{T_t f}$ and a sequence $t_n \downarrow 0$ such that $\widetilde{T_{t_n} f}(x) \to f(x)$, $n \to \infty$, q.e., for sufficiently many f. Going along a similar line as in the case of the Feller transition function, but ignoring successively the sets of capacity zero on which things might go wrong, we finally get a Hunt process outside some set of capacity zero.

§6.1 Construction of a Markovian Transition Function

Let the pair (X, m) be as in §1.1 and $\mathscr{E}$ be a regular Dirichlet form on $L^2(X; m)$. By making use of the potential theory developed in Chapter 3 in connection with the form $\mathscr{E}$, we now construct an m-symmetric Hunt process on X whose Dirichlet form is the given one $\mathscr{E}$. As a preliminary step, we construct in this section a transition function on X with time parameter Q_+. Here we denote by Q and Q_+ the set of all rational numbers and positive rational numbers respectively.

Lemma 6.1.1. Let F be a set. Consider a countable subset G of F and a countable collection S of mappings from $F \times F$ into F. Then there exists a countable set H such that

a) $G \subset H \subset F$ b) $s(H \times H) \subset H, \ \forall s \in S.$

Proof. Construct a sequence $\{s_1, s_2, \ldots\}$ such that each $s \in S$ appears infinitely often in the sequence. Let $G_0 = G$, $G_{n+1} = G_n \cup s_{n+1}(G_n \times G_n)$, $n = 1, 2, \ldots$. Then the set $H = \bigcup_{n=1}^{\infty} G_n$ is a desired one. In fact, for any $x, y \in H$ and $s \in S$, we can find n such that $x, y \in G_n$ and $s = s_{n+1}$. Q.E.D.

Let $\{T_t, t > 0\}$ and $\{G_\alpha, \alpha > 0\}$ be the semigroup and resolvent on $L^2(X; m)$ associated with the form $\mathscr{E}$ (§1.4) and $\{e_A, A \in \theta_0\}$ be the 1-equilibrium potentials associated with $\mathscr{E}$ (§3.1). We use those notions and notations employed in Chapter 3.

(I) Markovian kernels $\{\tilde{p}_t, t \in Q_+\}$ and $\tilde{R}_1$.

Lemma 6.1.2. There exist a regular nest $\{F_k^0\}$ and Markovian kernels $\{\tilde{p}_t, t \in Q_+\}$ and $\tilde{R}_1$ on X satisfying the following conditions:
(i) $\tilde{p}_t(C_\infty(X)) \subset C_\infty(\{F_k^0\})$, $\tilde{R}_1(C_\infty(X)) \subset C_\infty(\{F_k^0\})$,
(ii) $\tilde{p}_t u$ and $\tilde{R}_1 u$ are quasi-continuous versions of $T_t u$ and $G_1 u$ respectively for any non-negative Borel $u \in L^2(X; m)$.

Proof. Let us first show that there exists a countable set $B_0 \subset \mathscr{F} \cap C_0(X)$ such that

(6.1.1) B_0 is dense in $C_0(X)$

(6.1.2) $u, v \in B_0, \quad a \in Q \Rightarrow |u| \in B_0, \quad u + v \in B_0, \quad au \in B_0.$

In order to construct B_0, let $\{u_1, u_2, \ldots\}$ be a countable dense subfamily of $C_0(X)$. For each k and n, choose a $u_{k,n} \in \mathscr{F} \cap C_0(X)$ such that $\|u_k - u_{k,n}\|_\infty \leq 1/n$. The set $G = \{u_{k,n}: k, n = 1, 2, \ldots\}$ is dense in $C_0(X)$. Define the mappings $s_0, s_1, s_2, \ldots$ from $\{\mathscr{F} \cap C_0(X)\}^2$ into $\mathscr{F} \cap C_0(X)$ by $s_0(u, v) = |u|$, $s_1(u, v) = u + v$ and $s_i(u, v) = a_i u$, $i = 2, 3, \ldots$, where $\{a_2, a_3, \ldots\} = Q$. Applying the preceding lemma to G and $S = \{s_0, s_1, s_2, \ldots\}$, we get a countable set B_0 satisfying (6.1.1) and (6.1.2). Let

(6.1.3) $H_0 = (\bigcup_{t \in Q+} T_t(B_0)) \cup G_1(B_0).$

Then H_0 is countable and $H_0 \subset \mathscr{F}$ in view of Lemma 1.3.3. Choose a quasi-continuous version $\tilde{u}$ (in the restricted sense) of each $u \in H_0$ and let $\tilde{H}_0 = \{\tilde{u}: u \in H_0\}$. By Theorem 3.1.2(i), there exists then a regular nest $\{F_k^0\}$ on X such that $\tilde{H}_0 \subset C_\infty(\{F_k^0\})$.

Now let $Y_0 = \bigcup_{k=1}^{\infty} F_k^0$ and fix $t \in Q_+$. We construct a Markovian kernel $\tilde{p}_t$ on X satisfying (i) and (ii). The kernel $\tilde{R}_1$ can be constructed in the same way. By Theorem 3.1.2 (ii),

(6.1.4) $\widetilde{T_t(u+v)}\ (x) = \widetilde{T_t u}(x) + \widetilde{T_t v}(x),\ x \in Y_0,\ u,\ v \in B_0,$

(6.1.5) $\widetilde{T_t\ (au)}\ (x) = \widetilde{aT_t u}(x),\ x \in Y_0,\ u \in B_0,\ a \in \mathbf{Q}.$

By the Markovian nature of T_t, we also have

(6.1.6) $0 \leqq u \leqq 1,\ u \in B_0 \Rightarrow 0 \leqq \widetilde{T_t u}(x) \leqq 1,\ \forall\ x \in Y_0.$

(6.1.4), (6.1.5) and (6.1.6) imply

(6.1.7) $|\widetilde{T_t u}(x)| \leqq \|u\|_\infty,\ x \in Y_0,\ u \in B_0.$

It is then easy to see that, for each $x \in Y_0$, there exists a unique positive linear functional l_x on $C_\infty(X)$ satisfying

(6.1.8) $l_x(u) = \widetilde{T_t u}(x),\ \ \forall u \in B_0,\ |l_x(u)| \leqq \|u\|_\infty,\ \ \forall u \in C_\infty(X).$

To see this, choose for any $u \in C_\infty(X)$ a sequence $u_n \in B_0$ such that $\|u_n - u\|_\infty \to 0$, $n \to \infty$. Then $\widetilde{T_t u_n}(x)$ is uniformly convergent on Y_0. Denote the limit by $l_x(u)$. l_x is positive because $l_x(|u|) = \lim_{n\to\infty} l_x(|u_n|) \geqq 0$. The linearity and the inequality in (6.1.8) are also clear. Note that $l_x(u)$ belongs to the space $C_\infty(\{F_k^0\})$ as a function of $x \in Y_0$ because $\widetilde{T_t u_n} \in C_\infty(\{F_k^0\})$.

l_x admits a positive measure $\tilde{p}_t(x,\cdot)$ on X such that $l_x(u) = \tilde{p}_t u(x)$ and $\tilde{p}_t(x, X) \leqq 1$, $x \in Y_0$. Extend $\tilde{p}_t$ by setting $\tilde{p}_t(x, A) = 0$, $\forall x \in X - Y_0$, $\forall A \in \mathscr{B}(X)$; then $\tilde{p}_t$ becomes a Markovian kernel on X satisfying the condition (i).

It now suffices to derive the property (ii) for $u \in C_0^+(X)$ because then, by the same argument as in the proof of Theorem 4.3.3, we can get (ii) for non-negative $u \in L^2(X; m)$. Choose $u_n \in B_0$ which is uniformly convergent to $u \in C_0^+(X)$. Take $w \in B_0$ such that $w \geqq u$. Then the functions $v_n = (0 \vee u_n) \wedge w$ belong to B_0 and converge to u uniformly and in L^2-sense as well. Since $\tilde{p}_t v_n$ is a quasi-continuous version of $T_t v_n$, a standard argument already adopted in the proof of Theorem 4.3.3 implies that $\tilde{p}_t u$ is a quasi-continuous version of $T_t u$. Q.E.D

(II) A regular nest $\{F_k\}$.

Let $\{A_n\}$ be a countable basis of open sets of X. We may assume that each A_n is relatively compact. Let

(6.1.9) $\theta_1 = \{A\colon A \text{ is a finite union of } A_n\text{'s}\}$;

then $\theta_1 \subset \theta_0$. For each $A \in \theta_1$, choose a quasi-continuous Borel version $\tilde{e}_A$ of the 1-equilibrium potential e_A such that $0 \leqq \tilde{e}_A(x) \leqq 1$, $^\forall x \in X$.

Denote by $\tilde{H}$ the smallest subfamily of $\mathscr{F} \cap \mathscr{B}_b(X)$ such that

(H.1) $\tilde{H} \supset B_0,\ \{\tilde{e}_A,\ A \in \theta_1\}$,

(H.2) $\tilde{p}_t(\tilde{H}) \subset \tilde{H},\ t \in Q_+,\ \tilde{R}_1(\tilde{H}) \subset \tilde{H}$,

(H.3) $\tilde{H}$ is an algebra over Q.

$\tilde{H}$ is then countable. To see this, consider the mappings from $\{\mathscr{F} \cap \mathscr{B}_b(X)\}^2$ into $\mathscr{F} \cap \mathscr{B}_b(X)$ defined by $s_1\colon (u, v) \to u + v$, $s_{2,a}\colon (u, v) \to au$, $s_3\colon (u, v) \to uv$, $s_{4,t}\colon (u, v) \to \tilde{p}_t u$, $s_5\colon (u, v) \to \tilde{R}_1 u$. It then suffices to apply Lemma 6.1.1 to $G = B_0 \cup \{\tilde{e}_A\colon A \in \theta_1\}$ and $S = \{s_1, s_{2,a}, s_3, s_{4,t}, s_5\colon a \in Q,\ t \in Q_+\}$.

Lemma 6.1.3. There exists a regular nest $\{F_k\}$ satisfying the following: If $Y_1 = \bigcup_{k=1}^{\infty} F_k$, then

(i) $\tilde{H} \subset C_\infty(\{F_k\}),\ F_k \subset F_k^0,\ k = 1,2, \ldots$

(ii) $\tilde{e}_A(x) = 1,\ x \in A \cap Y_1,\ A \in \theta_1$,

(iii) $^\exists t_k \subset Q_+,\ t_k \downarrow 0,\ \tilde{p}_{t_k}u(x) \to u(x),\ (1/t_k)(\tilde{R}_1 u(x) - e^{-t_k}\tilde{R}_1\tilde{p}_{t_k}u(x)) \to u(x),\ x \in Y_1,\ u \in \tilde{H}$,

(iv) $\tilde{p}_t\tilde{p}_s u(x) = \tilde{p}_{t+s}u(x),\ x \in Y_1,\ t, s \in Q_+,\ u \in \tilde{H}$,

(v) $\tilde{p}_t\tilde{R}_1 u(x) = \tilde{R}_1\tilde{p}_t u(x),\ e^{-t}\tilde{p}_t\tilde{R}_1 u(x) \leqq \tilde{R}_1 u(x),\ x \in Y_1,\ t \in Q_+,\ u \in \tilde{H}_+$,

(vi) $e^{-t}\tilde{p}_t\tilde{e}_A(x) \leqq \tilde{e}_A(x),\ x \in Y_1,\ t \in Q_+,\ A \in \theta_1$,

(vii) $0 \leqq \tilde{e}_A(x) \leqq 1,\ x \in Y_1,\ A \in \theta_1$,

(viii) $\tilde{e}_A(x) \leqq \tilde{e}_B(x),\ x \in Y_1,\ A, B \in \theta_1,\ A \subset B$.

Proof. By Lemma 3.1.4, $\tilde{e}_A(x) = 1$ q.e. on A. Since θ_1 is countable, we can find a set N_1 such that Cap $(N_1) = 0$ and

(6.1.10) $\tilde{e}_A(x) = 1,\ ^\forall x \in A - N_1,\ ^\forall A \in \theta_1$.

By Lemma 1.3.3 and Theorem 3.1.4, each $u \in \tilde{H}$ admits a sequence $t_k \in Q_+$ $t_k \downarrow 0$ such that (iii) holds for q.e. $x \in X$. Since $\tilde{H}$ is countable, $\{t_k\}$ can be selected independently of $u \in \tilde{H}$ and (iii) holds for every $u \in \tilde{H}$ and for every $x \in X - N_2$, N_2 being a suitable set of capacity zero.

By virtue of Theorem 3.1.2 (i), there exists a nest $\{F_k\}$ satisfying the condition (i) and $X - \bigcup_{k=1}^{\infty} F_k \supset N_1 \cup N_2$. Denote the m-regularization by

$\{F_k\}$ again. Obviously, (i), (ii), and (iii) are satisfied for this $\{F_k\}$. (iv) and (v) hold because they are valid m-a.e. (vi), (vii) and (viii) follow from Lemma 3.1.1 and Theorem 3.2.1. Q.E.D.

(III) A Markovian transition function p_t, $t \in Q_+$.

Lemma 6.1.4. (i) There exists a Borel set $Y_2 \subset Y_1$ with $\mathrm{Cap}(X\text{-}Y_2) = 0$, $\tilde{p}_t(x, X - Y_2) = 0$, $\forall x \in Y_2$, $\forall t \in Q_+$.
(ii) Let

$$(6.1.11) \quad p_t(x, A) = \begin{cases} \tilde{p}_t(x, A), & x \in Y_2, A \in \mathscr{B}(X) \\ 0 & x \in X - Y_2, A \in \mathscr{B}(X), \end{cases}$$

Then $\{p_t\}_{t \in Q+}$ is a Markovian transition function on $(X, \mathscr{B}(X))$; namely, each p_t is a Markovian kernel on $(X, \mathscr{B}(X))$ and property (1.4.10) is satisfied for every $t, s \in Q_+$.

Proof. (i) $T_t I_{X-Y_1}(x) = 0$ m-a.e. because $X - Y_1$ is m-negligible. Since $\tilde{p}_t(x, X - Y_1)$ is a quasi-continuous modification of $T_t I_{X-Y_1}$ by Lemma 6.1.2 (ii), there exists a Borel sets $Y_1^{(1)} \subset Y_1$ such that $\mathrm{Cap}(X - Y_1^{(1)}) = 0$, $\tilde{p}_t(x, X - Y_1) = 0$, $\forall x \in Y_1^{(1)}$, $\forall t \in Q_+$. In this way, we can find a sequence of Borel sets $Y_1 \supset Y_1^{(1)} \supset Y_1^{(2)} \ldots$ such that $\tilde{p}_t(x, X - Y_1^{(k)}) = 0$, $\forall x \in Y_1^{(k+1)}$, $\forall t \in Q_+$, $\mathrm{Cap}(X - Y_1^{(k+1)}) = 0$. $Y_2 = \bigcap_{k=1}^{\infty} Y_1^{(k)}$ then possesses property (i).
(ii) It suffices to show the semigroup property,

$$(6.1.12) \quad p_t p_s u(x) = p_{t+s} u(x), \ x \in X, \ t, s \in Q_+, \ u \in B_0.$$

Both sides vanish when $x \in X - Y_2$. When $x \in Y_2$, (i) implies $p_t p_s u(x) = \tilde{p}_t(p_s u)(x) = \tilde{p}_t(\tilde{p}_s u)(x)$, which is also equal to $\tilde{p}_{t+s} u(x) = p_{t+s} u(x)$ by virtue of the preceding lemma. Q.E.D.

§6.2 Construction of a Symmetric Hunt Process

Let $X_\Delta = X \cup \{\Delta\}$ be the one-point compactification of X. When X is already compact, Δ is adjoined as an isolated point. Extend the Markovian transition function p_t on $(X, \mathscr{B}(X))$ of Lemma 6.1.4 to $(X_\Delta, \mathscr{B}(X_\Delta))$ by

$$(6.2.1) \quad \begin{cases} p'_t(x, A) = p_t(x, A - \{\Delta\}) + (1 - p_t(x, X)) I_A(\Delta), \ x \in X, \\ p'_t(\Delta, A) = I_A(\Delta). \end{cases}$$

Then $\{p'_t, t \in Q_+\}$ is a Markovian transition function on $(X_\Delta, \mathscr{B}(X_\Delta))$ with $p'_t(x, X_\Delta) = 1$, $x \in X_\Delta$.

Consider the following objects:

(6.2.2) $\Omega_0 = (X_\Delta)^{Q+}$

(6.2.3) $X^0_t(\omega) = \omega(t)$, $\omega \in \Omega_0$, $t \in Q_+$

(6.2.4) $\mathscr{M} = \sigma\{X^0_s;\ s \in Q_+\}$, $\mathscr{M}^0_t = \sigma\{X^0_s;\ s \leqq t,\ s \in Q_+\}$, $t \in Q_+$.

By the Kolmogorov extension theorem, there exists a Markov process $\mathbf{M}_0 = \{\Omega_0, \mathscr{M}, \mathscr{M}^0_t, X^0_t, P_x\}_{x \in X_\Delta}$ with state space $(X, \mathscr{B}(X))$, time parameter Q_+, and transition function p'_t of (6.2.1). In other words, $\mathbf{M}_0$ enjoys the properties (M. 1) $\sim$ (M. 4) of §4.1 with S and $[0, \infty]$ being replaced by X and Q_+ respectively and $\mathbf{M}_0$ is related to p_t of Lemma 6.1.4 by

(6.2.5) $P_x(X^0_t \in A) = p_t(x, A)$, $t \in Q_+$, $x \in X$, $A \in \mathscr{B}(X)$.

It is then easy to see that the Borel set Y_2 of Lemma 6.1.4 is $\mathbf{M}_0$-invariant:

(6.2.6) $P_x(X^0_t \in Y_2 \cup \Delta, \forall t \in Q_+) = 1$, $x \in Y_2$.

Let us set for any $t \geqq 0$,

(6.2.7) $\mathscr{M}_t = \bigcap_{s \in Q_+, s > t} \mathscr{M}^0_s$, $\mathscr{M}'_t = \sigma(\mathscr{M}_t, \mathscr{N})$,

where $\mathscr{N} = \{\Gamma \in \mathscr{M}: p_x(\Gamma) = 0\ \forall x \in Y_2\}$. $\{\mathscr{M}_t\}_{t \geqq 0}$ and $\{\mathscr{M}'_t\}_{t \geqq 0}$ are right continuous. We further introduce the hitting time of a set $F \subset X$ by

(6.2.8) $\sigma^0_F(\omega) = \inf\{t \in Q_+: X^0_t \in F\}$.

By convention, we let $\inf \phi = \infty$ and $e^{-\infty} = 0$.

Lemma 6.2.1. For $A \in \theta_1$, the function $\tilde{e}_A$ specified in §6.1 (II) satisfies the following inequality:

(6.2.9) $E_x(\exp(-\sigma^0_A)) \leqq \tilde{e}_A(x)$, $x \in Y_2$.

Proof. Fix $x \in Y_2$ and let $Y^0_t(\omega) = e^{-t}\tilde{e}_A(X^0_t(\omega))$, $\omega \in \Omega_0$, $t \in Q_+$.

Then $(Y_t^0, \mathscr{M}_t^0, P_x)_{t\in Q+}$ is a non-negative bounded supermartingale. Indeed, since $0 \leqq \tilde{e}_A \leqq 1$ and $e^{-(t-s)}p_{t-s}\tilde{e}_A = e^{-(t-s)}\tilde{p}_{t-s}\tilde{e}_A \leqq \tilde{e}_A$ on Y_2 by Lemma 6.1.3, it follows from (6.2.6) and the Markov property of $\mathbf{M}_0$ that $0 \leqq Y_t^0 \leqq 1$ P_x–a.s. and $E_x(Y_t^0 | \mathscr{M}_s^0) = e^{-t}p_{t-s}\tilde{e}_A(X_s^0) \leqq e^{-s}\tilde{e}_A(X_s^0) = Y_s^0$ P_x–a.s., $t > s$, $t, s \in Q_+$.

Using Doob's optional sampling theorem and noting the identity $\tilde{e}_A(y) = 1$, $^\forall y \in A \cap Y_2$ (Lemma 6.1.3), we can now get the inequality (6.2.9) in the same way as the proof of Lemma 4.3.1. Q.E.D.

As easy consequences of Lemma 6.2.1, we have

Lemma 6.2.2. (i) For any $G \in \theta_0$ and any quasi-continuous version $\tilde{e}_G$ of the 1-equilibrium potential e_G,

$E_x(\exp(-\sigma_G^0)) \leqq \tilde{e}_G(x)$ q.e.

(ii) If $G_n \in \theta_0$, $G_n \downarrow$, $\mathrm{Cap}(G_n) \to 0$, then

$p_x(\lim_{n\to\infty} \sigma_{G_n}^0 = \infty) = 1$ q.e.

Proof. (i) Choose $A_n \in \theta_1$, $n = 1,2, \ldots$, which increase to $G \in \theta_0$. By Lemma 6.1.3 (viii), $\tilde{e}_{A_n}(x)$, $x \in Y_2$, is increasing in n. Let the limit be $\tilde{e}_G$; then the preceding Lemma implies that the desired inequality holds for every $x \in Y_2$. $\tilde{e}_G$ is a quasi-continuous version of e_G because of Theorem 3.1.4 and the identity $\mathscr{E}_1(e_G - e_{A_n}, e_G - e_{A_n}) = \mathrm{Cap}(G) - \mathrm{Cap}(A_n)$ which decreases to zero as $n \to \infty$. The inequality holds for any quasi-continuous version of e_G in view of Lemma 3.1.4. Q.E.D.

We can now proceed to our key lemma concerning the regularity of sample paths.

Lemma 6.2.3. There exists a Borel set $Y_3 \subset Y_2$ with $\mathrm{Cap}(X - Y_3) = 0$ and the following statements hold:

(i) Let $\Omega_{01} = \{\omega \in \Omega_0\colon \lim_{k\to\infty} \sigma_{X-F_k}^0 = \infty\}$, $\Omega_{02} = \{\omega \in \Omega_0\colon$ the path $(X_s^0(\omega))_{s\in Q+}$ possesses at every $t \geqq 0$ the left and right limits inside $Y_2 \cup \Delta\}$ and $\Omega_1 = \Omega_{01} \cap \Omega_{02}$. Then $P_x(\Omega_1) = 1$, $^\forall x \in Y_3$.

(ii) Set

$$(6.2.10)\quad X_t(\omega) = \lim_{s\in Q+, s\downarrow t} X_s^0(\omega),\ \omega \in \Omega_1,\ t \geqq 0;$$

then $P_x(X_t = X_t^0,\ ^\forall t \in Q_+) = 1$, $x \in Y_3$.

(iii) $P_x(X_0 = x) = 1$, $x \in Y_3$.

(iv) Let $R_t(\omega) = \{X_s(\omega)\colon 0 \leqq s \leqq t\}$, $\omega \in \Omega_1$, and $\Omega_2 = \{\omega \in \Omega_1\colon$

$R_t(\omega)$ is a compact subset of X whenever $X_t(\omega) \in X\}$; then $P_x(\Omega_2) = 1$, $x \in Y_3$.

Proof. (i) By virtue of the second assertion of the preceding lemma, we can find a Borel set $Y_3 \subset Y_2$ such that $\mathrm{Cap}(X - Y_3) = 0$ and $P_x(\Omega_{01}) = 1$, $\forall x \in Y_3$.

Let us show the inclusion

(6.2.11) $\Omega_{01} - \Omega_{02} \subset \bigcup_{u \in H+} \{\omega \in \Omega_0 \colon \{\tilde{R}_1 u(X_s^0)\}_{s \in Q+}$ does not possess the right or left limit at some $t \geqq 0\}$.

We first observe that the subspace $\tilde{R}_1(\tilde{H}_+)$ of $C_\infty(\{F_k\})$ separates the points of $Y_1 \cup \Delta$. Indeed, by assuming that a pair $x, y \in Y_1 \cup \Delta$ satisfies $\tilde{R}_1 u(x) = \tilde{R}_1 u(y)$, $\forall u \in \tilde{H}_+$, we have $\tilde{R}_1 u(x) - e^{-t_k}\tilde{R}_1 \tilde{p}_{t_k} u(x) = \tilde{R}_1 u(y) - e^{-t_k}\tilde{R}_1 \tilde{p}_{t_k} u(y)$, $\forall u \in \tilde{H}_+$, which means $u(x) = u(y)$, $\forall u \in \tilde{H}_+$, by Lemma 6.1.3. Therefore $x = y$. Take $\omega \in \Omega_{01} - \Omega_{02}$; then there exists $t \geqq 0$ such that either $\lim_{s \in Q+, s \downarrow t} X_s^0(\omega)$ or $\lim_{s \in Q+, s \uparrow t} X_s^0(\omega)$ does not exist. Suppose, for instance, that the former case happens. Note that t is strictly less than $\sigma^0_{X-F_k}$ for some k because $\omega \in \Omega_{01}$. Hence, we can find sequences $s_l, s'_l \in Q_+$ and points $x, y \in F_k \cup \Delta$, $x \neq y$, such that $s_l \downarrow t$, $s'_l \downarrow t$, $X^0_{s_l}(\omega) \to x$, $X^0_{s'_l}(\omega) \to y$. By the above observation, however, $\tilde{R}_1 u(x) \neq \tilde{R}_1 u(y)$, for some $u \in \tilde{H}_+$. Taking the continuity of $\tilde{R}_1 u|_{F_k \cup \Delta}$ into account, we conclude that $\{\tilde{R}_1 u(X_s^0)\}_{s \in Q+}$ does not possess the right limit at t. The inclusion (6.2.11) is proven.

Fix $x \in Y_2$ and $u \in \tilde{H}_+$. Then Lemma 6.1.3 and (6.2.6) imply that $E_x(e^{-t}\tilde{R}_1 u(X_t^0) \mid \mathscr{M}_s^0) = e^{-t} p_{t-s} \tilde{R}_1 u(X_s^0) = e^{-t}\tilde{p}_{t-s}\tilde{R}_1 u(X_s^0) \leqq e^{-t}\tilde{R}_1 u(X_s^0)$, P_x-a.s., $s < t$, $s, t \in Q_+$; namely, the process $\{e^{-s}\tilde{R}_1 u(X_s^0), \mathscr{M}_s^0, P_x\}_{s \in Q+}$ is a non-negative bounded supermartingale. Therefore, the P_x-measure of the event inside the braces of the right hand side of (6.2.11) vanishes[1)] and, consequently,

(6.2.12) $P_x(\Omega_{01} - \Omega_{02}) = 0, \ x \in Y_2.$

(ii) For $u \in C(X_\Delta)$, $v \in B_0$, $t \in Q_+$ and $x \in Y_2$, we have from Lemma 6.1.3 and (6.2.6) again that $E_x(u(X_t^0)v(X_t)) = \lim_{t_k \downarrow 0} E_x(u(X_t^0)v(X^0_{t+t_k})) = \lim_{t_k \downarrow 0} E_x(u(X_t^0)p_{t_k}v(X_t^0)) = E_x(u(X_t^0)v(X_t^0))$. By the monotone lemma, we extend this to $E_x(h(X_t^0, X_t)) = E_x(h(X_t^0, X_t^0))$, $h \in \mathscr{B}_b(X_\Delta \times X_\Delta)$. Take as h a bounded metric on X to obtain $P_x(X_t = X_t^0) = 1$.

---

1) Cf. P.A.Meyer [65; VI, T3].

(iii) Lemma 6.1.3 implies $E_x(u(X_0)) = \lim_{t_k \downarrow 0} p_{t_k} u(x) = u(x)$, $u \in B_0$, $x \in Y_2$.
(iv) Let K_n be compact sets increasing to X and u_n be functions in B_0 such that $0 \leqq u_n \leqq 1$, $u_n(x) > 0$, $\forall x \in K_n$. We see from Lemma 6.1.3 that the function $v = \sum_{n \geqq 0} 2^{-n} u_n (\in C_\infty^+(X))$ satisfies

$$(6.2.13) \quad \tilde{R}_1 v(x) > 0, \ \forall x \in Y_1, \ e^{-t}\tilde{p}_t \tilde{R}_1 v(x) \leqq \tilde{R}_1 v(x), \ \forall x \in Y_1.$$

In fact, each $\tilde{R}_1 u_n$ is strictly positive on $K_n \cap Y_1$. Moreover, $\tilde{R}_1 v \in C_\infty(\{F_k\})$ by Lemma 6.1.2 and hence

$$(6.2.14) \quad \Omega_1 - \Omega_2 = \bigcup_{t \in Q+} \{\omega \in \Omega_1 : \tilde{R}_1 v(X_t) > 0, \inf_{0 \leqq s \leqq t} \tilde{R}_1 v(X_s) = 0\}.$$

Set $Z_s^0(\omega) = e^{-s}\tilde{R}_1 v(X_s^0(\omega))$, $\omega \in \Omega_0$, $s \in Q_+$. Then the process $\{Z_s^0, \mathscr{M}_s^0, P_x\}_{s \in Q+}$ is a non-negative bounded supermartingale by virtue of (6.2.13) and (6.2.6). Hence, $\{Z_t, \mathscr{M}'_t, P_x\}_{t \geqq 0}$ is a right continuous non-negative bounded supermartingale where we set $Z_t(\omega) = \lim_{s \in Q+, s \downarrow t} Z_s^0(\omega)$ if the limit exists and $Z_t(\omega) = 0$ otherwise.

Now (6.2.14) means $\Omega_1 - \Omega_2 \subset \bigcup_{t \in Q+} \{\omega \in \Omega_0 : Z_t(\omega) > 0, \inf_{0 \leqq s \leqq t} Z_s(\omega) = 0\}$, but the P_x-measure of the event inside the braces is equal to zero.[1] Q.E.D.

Lemma 6.2.4. There exists a Borel set $Y \subset Y_3$ with $\mathrm{Cap}(X - Y) = 0$ satisfying the following property: the set $\Gamma = \{\omega \in \Omega_2 : \exists t > 0, X_t(\omega)$ or $X_{t-}(\omega) \in X - Y\}$ is contained in a set $\Gamma_0 \in \mathscr{M}$ such that $P_x(\Gamma_0) = 0$, $\forall x \in Y$.

Proof. Choose a decreasing sequence of open sets $G_n \supset X - Y_3$ such that $\mathrm{Cap}(G_n) \to 0$. Let
$\Gamma_3 = \{\omega \in \Omega_2 : \exists t \geqq 0, X_t(\omega)$ or $X_{t-}(\omega) \in X - Y_3\}$,
$\Gamma_3^0 = \{\omega \in \Omega_2 : \lim_{n \to \infty} \sigma_{G_n}^0(\omega) < \infty\}$.
Then $\Gamma_3 \subset \Gamma_3^0$ and, furthermore, Lemma 6.2.2 implies that $P_x(\Gamma_3^0)$ vanishes for every x in some Borel set $Y_4 \subset Y_3$ with $\mathrm{Cap}(X - Y_4) = 0$. Applying the same argument to Y_4 in place of Y_3, we get Γ_4, Γ_4^0 and Y_5. In this way we have sequences $Y_3 \supset Y_4 \supset \ldots$, $\Gamma_3 \subset \Gamma_4 \subset \ldots$ and $\Gamma_3^0 \subset \Gamma_4^0 \subset \ldots$ such that $\mathrm{Cap}(X - Y_k) = 0$, $\Gamma_k \subset \Gamma_k^0$, $p_x(\Gamma_k^0) = 0$, $\forall x \in Y_{k+1}$, $k =$

---
1) Cf. P.A.Meyer [65; VI, T15].

3,4, $\cdots$. Let $Y = \bigcap_{k=3}^{\infty} Y_k$; then Cap $(X - Y) = 0$ and $\Gamma = \bigcup_{k=3}^{\infty} \Gamma_k$.
The lemma holds for $\Gamma_0 = \bigcup_{k=3}^{\infty} \Gamma_k^0$. Q.E.D.

Now we set

$$\text{(6.2.15)} \quad \Omega = \{\omega \in \Omega_2 - \Gamma_0 : X_t(\omega) = X_t^0(\omega),\ {}^\forall t \in Q_+\}$$

and denote the restrictions of $\mathscr{M}$, $\mathscr{M}_t^0$, $\mathscr{M}_t$, $X_t^0(t \in Q_+)$, $X_t(t \geqq 0)$ and P_x $(x \in Y_\Delta)$ to the set Ω by the same notations again. Consider then the process

$$\text{(6.2.16)} \quad \mathbf{M}_Y = \{\Omega, \mathscr{M}, \mathscr{M}_t, X_t, P_x\}_{x \in Y_\Delta}.$$

Lemma 6.2.5. $\mathbf{M}_Y$ is a Hunt process on $(Y, \mathscr{B}(Y))$.

Proof. It suffices to verify that $\mathbf{M}_Y$ satisfies the conditions (**M.** 1), (**M.** 2), (**M.** 5), (**M.** 6) of §4.1 together with the strong Markov property (4.1.3) and the quasi-left continuity (4.1.4). Conditions (**M.** 1), (**M.** 2) and (**M.** 5) follow from Lemma 6.2.3. By virtue of Lemma 6.2.3 and Lemma 6.2.4, the sample path $t \rightsquigarrow X_t(\omega)$, $\omega \in \Omega$, satisfies the regularity condition (**M.** 6). In particular, Lemma 6.2.3 (iv) implies the second property in (**M.** 6), In view of (6.2.15), each $\omega \in \Omega$ is uniquely decided by its sample path $\{X_s(\omega), s \in [0, \infty)\}$ and, consequently, the translation operator θ_t on Ω is well defined.

It is clear that $\{\mathscr{M}_t\}_{t \geqq 0}$ is a right continuous admissible family. In proving the strong Markov property with respect to $\{\mathscr{M}_t\}_{t \geqq 0}$, it is therefore sufficient to derive the relation (4.1.3)$'$ or, equivalently, the identity

$$\text{(6.2.17)} \quad E_x(f(X_{\sigma+s}) \cdot I_\Lambda) = E_x(p_s^Y f(X_\sigma) \cdot I_\Lambda), \quad x \in Y, \quad s \geqq 0, \quad f \in C_\infty(X),$$

where p_s^Y denotes the transition function of $\mathbf{M}_Y$: $p_s^Y(x, E) = P_x(X_s \in E)$, $x \in Y$.

We have from Lemma 6.2.3 (ii),

$$\text{(6.2.18)} \quad p_s^Y f(x) = p_s f(x), \quad s \in Q_+, \quad x \in Y, \ f \in C_\infty(X),$$

and, hence, Lemma 6.1.2 implies the inclusion

$$\text{(6.2.19)} \quad p_s^Y(C_\infty(X)) \subset C_\infty(\{F_k\}\,;\, Y), \ s \in Q_+,$$

where $C_\infty(\{F_k\}\,;\,Y)$ denotes the restrictions to Y of functions in $C_\infty(\{F_k\})$. On the other hand, $\dot{\sigma}_k(\omega) = \inf\{t \geqq 0\colon X_t(\omega) \in Y - F_k\}$ satisfies

(6.2.20) $\lim_{k\to\infty} \dot\sigma_k(\omega) = \infty, \omega \in \Omega$

because of the identity $\dot\sigma_k(\omega) = \sigma^0_{K-F_k}(\omega)$. Combining (6.2.19) with (6.2.20), we see for every $\omega \in \Omega$, $s \in Q_+$ and $t \in [0, \infty) (= \bigcup_{k=1}^{\infty} [0, \dot\sigma_k(\omega)),$

(6.2.21) $$\lim_{t'\downarrow t} p_s^Y f(X_{t'}(\omega)) = p_s^Y f(X_t(\omega)),$$
$$\lim_{t'\uparrow t} p_s^Y f(X_{t'}(\omega)) = p_s^Y f(X_{t-}(\omega)), \quad f \in C_\infty(X),$$

Using (6.2.21) we can show the strong Markov property (6.2.17) by a standard argument. First, take $s \in Q_+$ and approximate the $\{\mathscr{M}_t\}$-stopping time σ from above by $\sigma_n = \dfrac{[2^n\sigma]+2}{2^n}$, $n = 1, 2, \ldots$. Since

$$\Lambda \cap \left\{\frac{k-1}{2^n} \leqq \sigma < \frac{k}{2^n}\right\} \in \mathscr{M}_{k/2^n} \subset \mathscr{M}^0_{(k+1)/2^n} \text{ for } \Lambda \in \mathscr{M}_\sigma,$$

the simple Markov property of the process $\mathbf{M}_0$ implies

$$E_x(f(x_{\sigma_n+s})\cdot I_\Lambda) = \sum_k E_x((X_{\{(k+1)/2^n\}+s})\cdot I_{\Lambda\cap\{(k-1)/2^n\leqq\sigma<k/2^n\}}) =$$
$$\sum_k E_x(E_{X^0_{(k+1)/2^n}}(f(X_s^0))I_{\Lambda\cap\{(k-1)/2^n\leqq\sigma<k/2^n\}}) = E_x(p_s^Y f(X_{\sigma_n})\cdot I_V).$$

Letting $n \to \infty$, we get from (6.2.21) the desired identity (6.2.17) for $s \in Q_+$. (6.2.17) then holds for any $s \geqq 0$ because $p_s^Y f(x)$ is right continuous in s.

It remains to show the quasi-left continuity (4.1.4), but this follows from the strong Markov property and (6.2.21). Note that (6.2.21) also holds for $f \in C(X_\Delta)$ because $p_s^Y f = p_s^Y g + f(\Delta)$ with $g(x) = f(x) - f(\Delta)$. Now let $\{\mathscr{M}_t\}$-stopping times σ_n increase to σ. We may assume $\sigma(\omega) < \infty$, $\omega \in \Omega$. By setting $Z(\omega) = \lim_{n\to\infty} X_{\sigma_n}$, we have for any $f, g \in C(X_\Delta)$, $x \in Y$,

$$E_x(f(Z)g(X_\sigma)) = \lim_{s\in Q_+, s\downarrow 0} \lim_{n\to\infty} E_x(f(X_{\sigma_n})g(X_{\sigma_n+s})) =$$
$$\lim_{s\in Q_+, s\downarrow 0} \lim_{n\to\infty} E_x(f(X_{\sigma_n})\, p_s^Y g(X_{\sigma_n})) = \lim_{s\in Q_+, s\downarrow 0} E_x(f(Z)p_s^Y g(Z)) = E_x(f(Z)g(Z))$$

from which we get $P_x(Z = X_\sigma) = 1$ in the same way as in the proof of Lemma 6.2.3 (ii). Q.E.D.

On account of (6.2.18) and Lemma 6.1.2,

(6.2.22) $p_t^Y f$ is a quasi-continuous version of $T_t f$

for any $f \in C_0(X)$ and $t \in Q_+$. In fact, (6.2.22) holds for any $t > 0$. To see this, choose $t_n \in Q_+$ which decreases to $t > 0$, then $\lim_{n\to\infty} p_{t_n}^Y f(x) = p_t^Y f(x)$, $x \in Y$. On the other hand, $T_{t_n} f = T_{t_n-t}(T_t f)$ is $\mathscr{E}_1$-convergent to $T_t f$ according to Lemma 1.3.3. Hence, (6.2.22) is valid for $t > 0$ by virtue of Theorem 3.1.4 (ii).

Now Theorem 4.1.3 enables us to extend the Hunt process $\mathbf{M}_Y$ of Lemma 6.2.5 to a Hunt process $\mathbf{M}$ on the whole space $(X, \mathscr{B}(X))$ in such a way that

(6 2.23) Y is $\mathbf{M}$-invariant and $\mathbf{M}_Y = \mathbf{M}|_Y$,

(6.2.24) each point $x \in X - Y$ is a trap with respect to $\mathbf{M}$.

Since the restriction of the transition function of $\mathbf{M}$ to $(Y, \mathscr{B}(Y))$ equals p_t^Y and the set $X - Y$ is of capacity zero, (6.2.22) means that the transition function of the Hunt process $\mathbf{M}$ is m-symmetric and that the given semigroup $\{T_t, t > 0\}$ is determined by this transition function. Thus, we have arrived at our main existence theorem:

Theorem 6.2.1. Given a regular Dirichlet form $\mathscr{E}$ on $L^2(X; m)$, there exists an m-symmetric Hunt process $\mathbf{M}$ on $(X, \mathscr{B}(X))$ whose Dirichlet form is the given one $\mathscr{E}$.

Recall Theorem 4.3.6 of §4.3 which concerns the uniqueness of such a Hunt process: the Hunt process $\mathbf{M}$ in Theorem 6.2.1 is unique up to the equivalence specified in §4.3. Recall also Theorem 4.3.3 which asserts that the transition function and the resolvent of the Hunt process of Theorem 6.2.1 are necessarily quasi-continuous.

Combining Theorem 6.2.1 with Theorem 4.5.1, we get the following existence theorem of diffusions:

Theorem 6.2.2. The following two conditions are equivalent to each other for a regular Dirichlet form $\mathscr{E}$ on $L^2(X; m)$:

(i) $\mathscr{E}$ possesses the local property.

(ii) There exists an m-symmetric diffusion process on $(X, \mathscr{B}(X))$ whose Dirichlet form is the given one $\mathscr{E}$.

Finally, let us consider the case that X is an Euclidean domain $D \subset R^n$ and m is Lebesgue measure. We say that a system $\{\nu_{ij}\}_{1 \le i,j \le n}$ of (signed) Radon measures on D is an *admissible system* if it satisfies (1.2.3) and if it makes the following form closable on $L^2(D)$:

$$(6.2.25) \quad \begin{cases} \mathscr{E}(u, v) = \sum_{i,j=1}^{n} \int_D \dfrac{\partial u(x)}{\partial x_i} \dfrac{\partial v(x)}{\partial x_j} \nu_{ij}(dx) \\ \mathscr{D}[\mathscr{E}] = C_0^\infty(D). \end{cases}$$

We are concerned with the family $\mathfrak{X}$ of symmetric (with respect to Lebesgue measure) diffusions on D specified by

$\mathfrak{X} = \{\mathbf{M}: \mathbf{M}$ is a symmetric diffusion on D with no killing inside D, the Dirichlet form on $L^2(D)$ of $\mathbf{M}$ possesses $C_0^\infty(D)$ as its core$\}$.

Denote by $\dot{\mathfrak{X}}$ the equivalence classes of $\mathfrak{X}$ in the sense of §4.3.

Theorem 6.2.3. There is a one-to-one correspondence between the family of the equivalence classes $\dot{\mathfrak{X}}$ of diffusions and the family of admissible systems of measures.

Proof. Given an admissible system, the smallest closed extension $\mathcal{E}$ of the form (6.2.25) is a regular Dirichlet form on $L^2(D)$ with the local property, as we have already seen in §2.1. Then $\mathcal{E}$ admits a symmetric diffusion $\mathbf{M}$ on D by virtue of Theorem 6.2.2. $\mathbf{M}$ is an element of $\mathfrak{X}$ by Theorem 4.5.3.

Conversely, Theorems 4.5.3 and 2.2.2 imply that any $\mathbf{M} \in \mathfrak{X}$ admits a Dirichlet form $\mathcal{E}$ which is expressible as (6.2.25) on its core $C_0^\infty(D)$. Q.E.D.

Theorem 6.2.3 reduces the study of the class $\mathfrak{X}$ of diffusions to the admissibility of $\{\nu_{ij}\}$; namely, to the closability question on the form (6.2.25). In §2.1, we already took up this question and exhibited several admissible systems $\{\nu_{ij}\}$. Typical examples are given by (2.1.11) and (2.1.12),[1)] in which ν_{ij} are not necessarily absolutely continuous with respect to Lebesgue measure.

1) See N.Ikeda-S.Watanabe [44], in which the diffusion associated with the form (2.1.12) is constructed by the method of skew product.

Notes

Chapter 1. Basic Theory of Dirichlet Forms

The notion of Dirichlet space was introduced by A. Beurling-J. Deny [12] as a function space which is continuously embedded into an L^1_{loc}-space and on which every normal contraction operates. Their article [12] revealed without proof, all essential substances of the theory of Dirichlet spaces, most of which were amplified and proven later by J. Deny [24], [25].

This book only treats symmetric Dirichlet forms. See T. Kato [54] and G. Stampacchia [81] for non-symmetric forms. See M. Ito [50], H. Kunita [58], [59], J. Bliedner [13], J. Elliott [32], A. Ancona [5], Y. LeJan [63] and references therein for non-symmetric Dirichlet forms.

S. Albeverio and R. Høegh Krohn [1] considered a class of Dirichlet forms when the underlying space X is an infinite dimensional space, e.g., the space of tempered distributions.

§1.1. The present notion of Markovity of a symmetric form was introduced in M. Fukushima [37] but it is a slight variant of the notion of the unit contraction.

§1.4. Theorems 1.4.1 and 1.4.2 are due to A. Beurling-J. Deny [12] and J. Deny [24], [25]. The notion of the Dirichlet form in the wide sense (Theorem 1.4.3) is sometimes useful (cf. [34]). A systematic treatment of the translation invariant Dirichlet form (Examples 1.4.1, 1.4.2 and 1.5.2) can be found in J. Deny [25]. See also C. Berg-G. Forst [10] and C. J. Preston [73] in this connection.

§1.5. The notion of the extended Dirichlet space was introduced by M. L. Silverstein [78]. He also associates an "extended Dirichlet space" with a recurrent semigroup. See also T. Kori [56] in this connection. Theorem 1.5.4 recovers a theorem of J. Deny [25; Chap. 5, Th. 2]. The space defined by the right hand side of (1.5.14) can be found in N. Aronszajn-K. T. Smith [6].

Chapter 2. The Scope of Dirichlet Forms

§2.1. (1. c) is due to K. Sato. See also S. Albeverio, R. Høegh Krohn and L. Streit [2] and M.L. Silverstein [79] for other important families of closable Markovian symmetric forms.

§2.2. See G. Allain [3] for a different proof of Theorem 2.2.1. Another proof of Theorem 2.2.2 can be found in L.-E. Andersson [4].

§2.3. Theorem 2.3.1 is due to T. Watanabe (cf. [34; Th. 5.1]). This book only specifies the minimum and maximum elements of $\mathscr{A}_M(S)$ for $S = 1/2\Delta$. In M. Fukushima [34], the family $\mathscr{A}_M(S)$ is characterized by means of a class of Dirichlet forms in the wide sense, living on the Martin boundary. See Examples 1.2.3 and 5.5.2 in this connection. This kind of characterization has been generalized (to more general S's) by H. Kunita [59], J. Elliott [31], M. L. Silverstein [77], [78], [79], [80], and Y. LeJan [62], [63]. These works may be classified as the boundary theory for Dirichlet spaces.

Chapter 3. Potential Theory of Dirichlet Forms

§3.1. Theorems 3.1.1, 3.1.3, and 3.1.4 were first proven by J. Deny (cf. J. Deny [23], J. Deny-J. L. Lions [26]). Theorem 3.1.2 and Theorem 3.1.5 are taken from M. Fukushima [37] and M. L. Silverstein [78], respectively.

The sets of zero capacity have played important roles as exceptional sets in the Fourier series and boundary limit theorems. See A. Beurling [11], N. K. Bary [8], A. Zygmund [88], L. Carleson [16] and C. J. Preston [73] in this connection.

§3.2. Theorem 3.2.2 was presented in A. Beurling-J. Deny [12] and proven in J. Deny [25]. Theorem 3.2.3 is essentially due to M. L. Silverstein [78], in which only positive Radon measures charging no set of zero capacity are considered.

§3.3. Lemma 3.3.2 and Theorem 3.3.4 are taken from M. L. Silverstein [78] and J. Deny [25], respectively.

Chapter 4. Analysis in Symmetric Markov Processes

The first two chapters of R. M. Blumenthal-R. K. Getoor [14] are frequently referred to in §4.1 and §4.2. P. A. Meyer [65] is referred to in §4.3 and §4.4 for the optional sampling theorem of supermartingales. But the descriptions in Chapter 4 are otherwise, rather self-contained.

Theorem 4.2.3 remains true for a non-symmetric Hunt process provided that it admits a non-symmetric Dirichlet form (*Y.* LeJan [62]). M. Kanda [53] showed that every semi-polar set is still polar for a multidimensional non-symmetric Cauchy process which does not admit a (non-symmetric) Dirichlet form (cf. C. Berg-G. Forst [9]).

Theorems 4.3.1 and 4.3.2 were first shown for Brownian motions by S. Kakutani [51], J. L. Doob [27], and J. Deny-J. L. Lions [26]. See J. L.

Doob [28] and Example 4.3.1 in this connection. Theorems 4.3.1 and 4.3.4 have been extended to any symmetric Hunt process without assuming the regularity of the associated Dirichlet space by using the notion of fine capacity (cf. [39]).

Theorem 4.3.1 legitimates our specific use of the term "exceptional", which was called "essentially polar" in S. C. Port-C. J. Stone [72] (for spatially homogeneous Markov processes) and "almost polar" in M. Fukushima [38] (for standard processes possessing dual processes). M. Fukushima [38] and M. Sur [83] have shown that the "true" exceptional set in the Chacon-Ornstein ergodic theorem is actually an almost polar set.

There are several methods of solving the classical Dirichlet problem for harmonic functions. Theorem 4.4.1 identifies a probabilistic method due to S. Kakutani [52] with a method of projection due to H. Weyl [85]. See K. Ito [47] and E. B. Dynkin [30] for the probabilistic principle.

The use of supermartingales in the proof of Lemma 4.3.1 and Theorem 4.4.1 is due to M. L. Silverstein [78]. Lemma 4.5.2 and the notion of the resurrected Dirichlet form are also taken from [78]. The method of proof of Theorem 4.5.1 is taken from P. Courrège-P. Priouret [19].

Most theorems in §4.3, §4.4, and §4.5 have been extended to Hunt processes associated with non-symmetric Dirichlet forms by S. Carrillo Menendez [17], Y. LeJan [62], [63], and M. L. Silverstein [80].

Chapter 5. Further Analysis by Additive Functionals

The contents of Chapter 5 are self-contained except that P. A. Meyer [67; III] is referred to in §5.2 and §5.4 for some basic properties of MAF's and that R. M. Blumenthal-R. K. Getoor [14; Chap. V] is referred to in §5.5 for the support of a PCAF.

The class of PCAF's (admitting no exceptional set N) was characterized in terms of a certain class of Borel measures by P. H. McKean–H. Tanaka [64], by A. D. Wentzell [86] for multidimensional Brownian motion and by D. Revuz [75] for the Hunt process satisfying some condition of duality. This class of measures is narrower and less simple than the class S of smooth measures in the present sense. See also E. B. Dynkin [30] in this connection.

Formula (5.1.9) was used explicitly by D. Revuz [75] in specifying the correspondence. The method in the proof of Theorems 5.1.1 and 5.1.2 (resp. Theorem 5.1.3) is quite similar to that in [64] (resp. [75]).

As an application of Theorem 5.1.1, H. Nagai [71] relates the optimal stopping of a difference of PCAF's to a certain variational inequality involving the associated measure.

M. L. Silverstein [78] introduced the notion of energy measure of a function in $\mathscr{F}$ by means of the formula (5.4.1). Lemma 5.4.1 and Theorem 5.4.1 are due to Y. LeJan [63]. They used the energy measures in developing the boundary theory for Dirichlet spaces.

The stochastic integral based on an MAF (admitting no exceptional set) was first introduced by M. Motoo-S. Watanabe [70]. The method was then combined with the Doob-Meyer decomposition theorem for supermartingales [65] and extended to square integrable martingales by H. Kunita-S. Watanabe [60]. Ito's formula was also established for semimartingales. See a fine exposition [67] of P. A. Meyer in this connection. The contents of §5.2, §5.3, and §5.4 may be viewed as a continuation of such development in the realm of Dirichlet spaces.

See N. Ikeda-S. Watanabe [45] for an exhaustive account of stochastic calculus. Stochastic calculus was used in N. Ikeda-S. Watanabe [44] to get the expressions of a certain class of Dirichlet forms.

Theorem 5.5.1 is due to M. L. Silverstein [78]. This is a starting point in a probabilistic approach to the boundary theory for Dirichlet spaces.

Chapter 6. Construction of Symmetric Markov Processes

A Hunt process associated with a regular Dirichlet form was first constructed in M. Fukushima [36]. The present method of construction is due to M. L. Silverstein [78]. A variant of Lemma 6.1.1 was also used in constructing Ray's resolvents (cf. F. Knight [55] and H. Kunita-T. Watanabe [61]). Our case differs from the case of the Feller semigroup mainly in that we need an additional inequality (6.2.9).

The results of this chapter have been extended to non-symmetric Dirichlet forms by S. Carrillo Menendez [17].

In connection with Theorem 6.2.3, it would be interesting to ask if an admissible system $\{\nu_{ij}\}$ admits a Feller semigroup. When ν_{ij} are absolutely continuous and the density functions satisfy certain conditions, the theory of elliptic and parabolic differential equations have afforded affirmative answers (cf. E. B. Dynkin [30] and H. Kunita [59]).

Bibliography

[1] S. Albeverio and R. Høegh Krohn, Dirichlet forms and diffusion processes on rigged Hilbert spaces, *Z. Wahrscheinlichkeitstheorie verw. Gebiete*, **40** (1977), 1–57.

[2] S. Albeverio, R. Høegh Krohn and L. Streit, Energy forms, Hamiltonians, and distorted Brownian paths, *J. Math. Phys.*, **18** (1977), 907–917.

[3] G. Allain, Sur la représentation des forms de Dirichlet, *Ann. Inst. Fourier*, **25**, 3 et 4 (1975), 1–10.

[4] L.-E. Andersson. On the representation of Dirichlet forms, *Ann. Inst. Fourier*, **25**, 3 et 4 (1975), 11–25.

[5] A. Ancona, Continuité des contractions dans les espaces de Dirichlet, Séminaire de Theorie du Potentiel Paris, No. 2, *Lecture Notes in Math.*, vol. 563, Springer-Verlag, Berlin-Heidelberg-New York, 1976.

[6] N. Aronszajn and K. T. Smith, Functional spaces and functional completion, *Ann. Inst. Fourier*, **6** (1956), 125–185.

[7] ———, Theory of Bessel potentials. Part I, *Ann. Inst. Fourier*, **11** (1961), 385–475.

[8] N. K. Bary, "*A treatise on trigonometric series*", Pergamon, Oxford, 1964.

[9] C. Berg and G. Forst, Non-symmetric translation invariant Dirichlet forms, *Invent. Math.*, **21** (1973), 199–212.

[10] ———, "*Potential theory on locally compact abelian groups*", Springer-Verlag, Berlin-Heidelberg-New York, 1975.

[11] A. Beurling, Ensemble exceptionnels, *Acta Math.*, **72** (1939), 1–13.

[12] A. Beurling and J. Deny, Dirichlet spaces, *Proc. Nat. Acad. Sci. U.S.A.*, **45** (1959), 208–215.

[13] J. Bliedner, Functional spaces and their exceptional sets, Seminar on Potential Theory II, *Lecture Notes in Math.*, vol. 226, Springer-Verlag, Berlin-Heidelberg-New York, 1971.

[14] R. M. Blumenthal and R. K. Getoor, "*Markov processes and Potential theory*," Academic Press, New York and London, 1968.

[15] M. Brelot, "*Éléments de la théorie classique du potentiel*", *Les Cours de Sorbonne*, Centre de documentation Universitaire, 1959.

[16] L. Carleson, "*Selected problems on exceptional sets*", Van Nostrand, Princeton, 1967.

[17] S. Carrillo Menendez, Processus de Markov associé a une forme de Dirichlet non symétrique, *Z. Wahrscheinlichkeitstheorie vew. Gebiete*, **33** (1975), 139–154.

[18] R. Courant and D. Hilbert, "*Methods of Mathematical Physics,* vol. **1**, Wiley (Interscience), New York, 1953.

[19] P. Courrège and P. Priouret, Axiomatique du problème de Dirichlet de processus de Markov, Séminaire Brelot-Choquet-Deny, 8e année, Paris, 1963–64.

[20] C. Dellacherie, "*Capacités et processus stochastiques*", Springer-Verlag, Berlin-Heidelberg-New York, 1972.

[21] C. Dellacherie and P. A. Meyer, "*Probabilités et potentiel*", *Hermann*, Paris, 1975.
[22] J. Deny, Les potentiels d'énergie finie, *Acta Math.*, **82** (1950), 107–183.
[23] ———, Théorie de la capacité dans les espaces fonctionnels, Séminaire Brelot-Choquet-Deny, 9e année, Paris, 1964–65.
[24] ———, Principe complet du maximum et contractions, *Ann. Inst. Fourier*, **15**, 1 (1965), 259–272.
[25] ———, Méthodes Hilbertiennes et théorie du potentiel, Potential Theory, Centro Internazionale Mathematico Estivo, Edizioni Cremonese, Roma, 1970.
[26] J. Deny and J. L. Lions, Les espaces du type de Beppo Levi, *Ann. Inst. Fourier*, **5** (1953/54), 305–370.
[27] J. L. Doob, Semimartingales and subharmonic functions, *Trans. Amer. Math. Soc.*, **77** (1954), 86–121.
[28] ———, Boundary properties of functions with finite Dirichlet integrals, *Ann. Inst. Fourier*, **12** (1962), 573–621.
[29] ———, Applications to analysis of a topological definition of smallness of a set, *Bull. Amer. Math., Soc.*, **72** (1966), 579–600.
[30] E. B. Dynkin, "*Markov processes*" Springer-Verlag, Berlin-Heidelberg-New York, 1965.
[31] J. Elliott, Dirichlet spaces and boundary conditions for submarkovian resolvents, *J. Math. Anal. App.*, **36** (1971), 251–282.
[32] ———, On unsymmetric Dirichlet forms, *Can. J. Math.*, **25** (1973), 252–260.
[33] W. Feller, Generalized second order differential operators and their lateral conditions, *Illinois J. Math.*, **1** (1957), 459–504.
[34] M. Fukushima, On boundary conditions for multi-dimensional Brownian motions with symmetric resolvent densities, *J. Math. Soc. Japan*, **21** (1969), 58–93.
[35] ———, Regular representations of Dirichlet spaces, *Trans. Amer. Math. Soc.*, **155** (1971), 455–473.
[36] ———, Dirichlet spaces and strong Markov processes, *Trans. Amer. Math. Soc.*, **162** (1971), 185–224.
[37] ———, On the generation of Markov processes by symmetric forms, Proc. 2nd Japan-USSR symp. on probability theory, *Lecture Notes in Math.*, vol. 330, Springer-Verlag, Berlin-Heidelberg-New York, 1973.
[38] ———, Almost polar sets and an ergodic theorem, *J. Math. Soc. Japan*, **26** (1974), 17–32.
[39] ———, Potential theory of symmetric Markov processes and its applications, Proc. 3rd Japan-USSR symp. on probability theory, *Lecture Notes in Math.*, vol. 550, Springer-Verlag, Berlin-Heidelberg-New York, 1976.
[40] ———, On additive functionals admitting exceptional sets, *J. Math. Kyoto Univ.*, **19** (1979), 191–202.
[41] ———, A decomposition of additive functionals of finite energy, *Nagoya Math. J.* .**74** (1979), 137–168.
[42] M. Hamza, Détermination des formes de Dirichlet sur R^n, Thèse 3e cycle, Orsay, 1975.
[43] G. A. Hunt, Markoff processes and potentials I, II, and III, *Illinois J. Math.*, **1** (1957), 44–93; **1**(1957) 316–369; **2** (1958), 151–213.
[44] N. Ikeda and S. Watanabe, The local structure of a class of diffusions and

related problems, Proc. 2nd Japan-USSR symp. on probability theory, *Lecture Notes in Math.*, vol. 330, Springer-Verlag, Berlin-Heidelberg-New York, 1973.

[45] ———, "*Stochastic differential equations and diffusion processes*", Kodansha Scientific, Tokyo, (in press).

[46] K. Ito, Stochastic integral, *Proc. Imp. Acad. Tokyo*, **20** (1944), 519–524.

[47] ———, "*Lectures on stochastic processes*", Tata Institute of Fundamental Research, Bombay, 1960.

[48] ———, "Stochastic processes", *Lecture Notes Series*, No. 16, Aarhus Universitet, Matematisk Institut, 1969.

[49] K. Ito and H. P. McKean, "*Diffusion processes and their sample paths*", Springer-Verlag, Berlin-Heidelberg-New York, 1965.

[50] M. Ito, A note on extended regular functional spaces, *Proc. Jap. Acad.*, **43** (1967), 435–440.

[51] S. Kakutani, On Brownian motion in *n*-space, *Proc. Imp. Acad. Tokyo*, **20** (1944), 648–652.

[52] ———, Two-dimensional Brownian motion and harmonic functions, *Proc. Imp. Acad. Tokyo*, **20** (1944), 706–714.

[53] M. Kanda, Characterization of semipolar sets for processes with stationary independent increments, *Z. Wahrscheinlichkeitstheorie verw. Gebiete*, **42** (1978), 141–154.

[54] T. Kato, "*Perturbation theory for linear operators*", Springer-Verlag, Berlin-Heidelberg-New York, 1966.

[55] F. Knight, Note on regularization of Markov processes, *Illinois J. Math.*, **9** (1965), 548–552.

[56] T. Kori, La théorie des fonctionnels a nullité 1 et le probleme de Neumann sur les espaces harmoniques, *Ann. Inst. Fourier*, **27**, 4 (1977), 45–119.

[57] M. G. Krein, The theory of self-adjoint extensions of semi-bounded Hermitian transformations and their applications, Part I, *Mat. Sbornik*, **20** (1947), 431–495.

[58] H. Kunita, Sub-Markov semi-groups in Banach lattices, Proc. international conference on functional analysis and related topics, 1969. Tokyo.

[59] ———, General boundary conditions for multi-dimensional diffusion processes, *J. Math. Kyoto Univ.*, **10** (1970), 273–335.

[60] H. Kunita and S. Watanabe, On square integrable martingales, *Nagoya Math. J.*, **30** (1967), 209–245.

[61] H. Kunita and T. Watanabe, Some theorems concerning resolvents over locally compact spaces, Proc. Fifth Berkeley Symp. Math. Statist. and Probability vol. II, Berkeley, 1967.

[62] Y. LeJan, Balayage et formes de Dirichlet, *Z. Wahrscheinlichkeitstheorie verw. Gebiete*, **37** (1977), 297–319.

[63] ———, Mesures associées a une forme de Dirichlet. Applications, *Bull. Soc. Math. France*, **106** (1978), 61–112.

[64] H. P. McKean and H. Tanaka, Additive functionals of the Brownian path., *Memoirs Coll. Sci. Univ. Kyoto, A. Math.*, **33** (1961), 479–506.

[65] P. A. Meyer, "*Probability and Potentials*", Ginn Blaisdell, Massachusetts, 1966.

[66] ———, "Processus de Markov", *Lecture Notes in Math.*, vol. 26, Springer-Verlag, Berlin-Heidelberg-New York, 1967.

[67] ———, Intégrales stochastiques I, II, III, Séminaire de Probabilités I, *Lecture Notes in Math.*, vol. 39, Springer-Verlag, Berlin-Heidelberg-New York, 1967.

[68] ———, Martingales locales fonctionneles additives I, Séminaire de Probabilités XII, *Lecture Notes in Math.*, vol. 649, Springer-Verlag, Berlin-Heidelberg-New York, 1978.

[69] S. Mizohata, "*The theory of partial differential equations*", Cambridge University Press, London, 1973.

[70] M. Motoo and S. Watanabe, On a class of additive functionals of Markov processes, *J. Math. Kyoto Univ.*, **4** (1965), 429–469.

[71] H. Nagai, On an optimal stopping problem and a variational inequality, *J. Math. Soc. Japan*, **30** (1978), 303–312.

[72] S. C. Port and C. J. Stone, Infinitely divisible processes and their potential theory, *Ann. Inst. Fourier*, **21**, 2 (1971), 157–275.

[73] C. J. Preston, A theory of capacities and its application to some convergence results, *Advances in Math.*, **6** (1971), 78–106.

[74] D. Ray, Resolvents, transition functions, and strongly Markovian processes, *Ann. Math.*, **70** (1959), 43–72.

[75] D. Revuz, Mesures associées aux fonctionnelles additives de Markov I, *Trans. Amer. Math. Soc.*, **148** (1970), 501–531.

[76] F. Riesz and B. Sz. Nagy, "*Functional analysis*", Unger, New York, 1955.

[77] M. L. Silverstein, Classification of stable symmetric Markov chains, *Indiana J. Math.*, **24** (1974), 29–77.

[78] ———, "Symmetric Markov processes," *Lecture Notes in Math.*, vol. 426, Springer-Verlag, Berlin-Heidelberg-New York, 1974.

[79] ———, "Boundary theory for symmetric Markov processes," *Lecture Notes in Math.*, vol. 516, Springer-Verlag, Berlin-Heidelberg-New York, 1976.

[80] ———, Application of the sector condition to the classification of sub-markovian semigroups, *Trans. Amer. Math. Soc.*, **244** (1978), 103–146.

[81] G. Stampacchia, Forms bilineaires coercitives sur les ensemble convexes, *C. R. Acad. Sc. Paris*, **258** (1964), 4413–4416.

[82] E. M. Stein, "*Singular integrals and differentiability properties of functions*", Princeton University Press, Princeton, 1970.

[83] M. Sur, An ergodic theorem for Markov processes I. II, *Theory Prob. Applications*, **21** (1976), 400–406; **22** (1977), 692–707.

[84] S. Watanabe, On discontinuous additive functionals and Lévy measures of Markov process, *Japanese J. Math.*, **34** (1964), 53–70.

[85] H. Weyl, The method of orthogonal projection in potential theory, *Duke Math. J.*, **7** (1940), 411–444.

[86] A. D. Wentzell, nonnegative additive functionals of Markov processes, *Dokl. Akad. Nauk SSSR*, **137** (1961), 17–20.

[87] K. Yosida, "*Functional analysis*", Springer-Verlag, Berlin-Heidelberg-New York, 1968.

[88] A. Zygmund, "*Trigonometric series*", Cambridge,1959.

Index